RAND NATIONAL DEFENSE RESEARCH INSTITUTE

Building Toward an Unmanned Air System Training Strategy

Bernard D. Rostker, Charles Nemfakos, Henry A. Leonard, Elliot Axelband, Abby Doll, Kimberly N. Hale, Brian McInnis, Richard Mesic, Daniel Tremblay, Roland J. Yardley, Stephanie Young

Prepared for OUSD (P&R), Directorate for Training and Readiness Policy and Programs

For more information on this publication, visit www.rand.org/pubs/research_reports/RR440.html

Library of Congress Cataloging-in-Publication Data

Rostker, Bernard.
Building toward an unmanned aircraft system training strategy / Bernard D. Rostker [and ten others].
pages cm
Includes bibliographical references.
ISBN 978-0-8330-8531-3 (pbk. : alk. paper)
1. Drone aircraft—United States. 2. Drone aircraft pilots—Training of—United States.
I.Title.

UG1242.D7R64 2014
358.4'156—dc23 2014025166

Published by the RAND Corporation, Santa Monica, Calif.

Preface

During the most recent rounds of Base Closure and Realignment Commission activities in 2005, a significant number of training bases were closed. In light of the introduction of new technologies and the great expansion of unmanned aircraft systems (UASs) in the force, Department of Defense planners and some in Congress have become concerned that the existing training infrastructure—bases and their training support facilities—may not be adequate to train UAS air and ground components and the ground forces that use such equipment to capitalize fully on their capabilities. Accordingly, the Deputy Director, Readiness and Training Policy and Programs in the Office of the Under Secretary of Defense for Personnel and Readiness (OUSD [P&R]) asked the RAND Corporation to assess the adequacy of UAS training to support current and future requirements. In addition, the House Armed Services Committee report accompanying the Fiscal Year 2013 National Defense Appropriations Act raised a number of questions concerning training strategies with particular reference to the use of simulators to facilitate training.[1]

This report considers three issues: (1) the development of a general concept for UAS training in the context of current and anticipated future UAS inventories, (2) the development of an appropriate framework based on the general concept to address UAS training requirements, including the appropriate use of simulators, and (3) the airspace requirements necessary for UAS training. The research reported on here covers UASs in the Army, Navy, Air Force, and Marine Corps as fielded and plans as they existed during 2012. A RAND team carried out extensive field visits to understand the current ability of the services to conduct (1) service-specific training and (2) joint training at both home station and joint training facilities. The research will be of interest to those concerned with UAS programs and operations; joint training; or more generally, the incorporation of disruptive technologies into Department of Defense programs and operations.

This research was sponsored by the OUSD (P&R) Directorate for Training and Readiness Policy and Programs, in coordination with the Joint Forces Command's Joint Warfighting Center and the Joint Unmanned Aircraft Systems Center of Excellence. The research was conducted within the Forces and Resources Policy Center of the RAND National Defense Research Institute (NDRI), a federally funded research and development center sponsored by the Office of the Secretary of Defense, the Joint Staff, the Unified Combatant Commands, the Navy, the Marine Corps, the defense agencies, and the Defense Intelligence Community. For

[1] House Armed Services Committee, *Conference Report Accompanying the Fiscal Year 2013 National Defense Appropriations Act*, 112th Cong., 2nd Sess., December 17, 2012, p. 136.

more information on the RAND Forces and Resources Policy Center, see http://www.rand.org/nsrd/ndri/centers/frp.html or contact the director (contact information is provided on the web page).

Contents

Figures and Tables

Figures

Tables

Summary

Unmanned aircraft systems (UASs) have become increasingly prevalent in and important to U.S. military operations. The number of systems has surged, as has the slice of the defense budget allocated to them. Roles have also changed. Initially employed only as reconnaissance or intelligence platforms, they now carry out other missions, including attacking enemy forces. Successful operational tests and demonstrations of the expanded range of UAS capabilities have led to rapid fielding of new systems, often placing unanticipated demands on logistics and training systems and on field commanders to employ new systems effectively. UASs must now be integrated into the training programs of the services. Building a responsive, effective, and efficient UAS training program is a challenge during times of reduced budgets. Any new program must be based on a review of existing training capabilities and new investments across the services.

Focus of this Research

The Deputy Director, Readiness and Training Policy and Programs, in the Office of the Under Secretary of Defense for Personnel and Readiness (OUSD [P&R]) asked the RAND National Defense Research Institute (NDRI) to assess the adequacy of UAS training to support current and future requirements. Proposals to resolve service and joint UAS training issues must be informed by a clear understanding of current problems, opportunities for correction, and associated costs and benefits of the corrections. In this report, we address a number of issues, including (1) a general concept for UAS training; (2) an appropriate framework to address UAS training requirements, including the use of simulators; and (3) the limitations of the infrastructure to accommodate UAS training, including those due to national airspace restrictions. The research considers UASs in the Army, Navy, Air Force, and Marine Corps, both those that are currently fielded and those that were planned to be fielded as of 2012.

Disruptive Technologies and Acquisition Processes

In 1995, Clayton Christensen introduced the term *disruptive technology*, distinguishing it from what he termed *sustaining technology*. In his nomenclature, a sustaining technology improves the performance of an existing system and does not require significant structural adjustments to processes, organizations, or operational paradigms. Disruptive technologies, however, change the way a business operates. The UAS is a disruptive technology.

UASs found their way into the force by means of rapid acquisition procedures that were designed to deliver new technologies to the warfighter much faster than traditional procedures could, but often at the cost of not having in place the support these new systems required to become a sustained part of the force structure. Because of the rapid acquisition of these systems, the services have had to rely upon contractor support in ways that are inconsistent with the full and sustaining integration of these systems into service inventories. Such integration starts with the development of an appropriate concept for training.

Training Concepts and Frameworks

We will begin by discussing the general ideas that should guide UAS training—the training *concept*. Then, we will consider how the parts identified in the concept fit together into a conceptual structure for that training—a training *framework* designed to enhance combat power and other operational capabilities.

Joint Operations; Multiservice Tactics, Techniques, and Procedures; and UAS Training

Military operations today are usually referred to as *joint operations*, emphasizing the interdependence of the services. That might lead one to think that joint doctrine; joint concepts of operation; and joint tactics, techniques, and procedures would provide focus for training. Clearly, training does not take place for its own sake; it should be driven by doctrine. In his cover letter to Joint Publication 3-0, 2011, ADM Mike Mullen stated that it established "the framework for our forces' ability to fight as a joint team," but nowhere does it discuss the management of UASs. Moreover, nowhere in any other joint publication is there a grand doctrine for the employment or management of UASs on the battlefield. While the services have agreed to a set of multiservice tactics, techniques, and procedures (MTTPs) to be incorporated into their respective training programs, note the use of *multiservice* rather than *joint*.

At least for the foreseeable future, each service will continue to field UASs and must bring to the fight forces fully capable of operating them and integrating their capabilities into operations. While grand doctrinal issues remain unresolved, common procedures agreed to and trained by the services will remain the foundation on which American military operations must be built. In this way, rather than being the foundation for UAS training, joint operations are the goal of such training, which must be firmly grounded on the UAS training each service gives both its supported and supporting units.

The general concept we advance for UAS training here rests on the notion that the greatest beneficial effect—that is, an increase in the force's operational effectiveness—occurs with a synergy among the UAS platforms, those who operate them, and those who integrate them into combat operations. While the report considers training concepts for ground, air, and sea forces, they share a common framework. Figure S.1 depicts our framework for ground and air forces; the one for naval forces is similar.

The concept underpinning the framework calls for increasing integration of its several parts as the level of training moves up the pyramid. Army systems appear on the left, Air Force on the right. At the bottom, training begins with individual training for operators and leaders, grounding each group in the basics of the system. Training then progresses to the small unit level, which, as with individual training, is largely done at home station. Moving up the pyramid, training occurs at increasingly higher echelons, battalion and brigade. Moving to joint

Figure S.1
UAS Training Framework

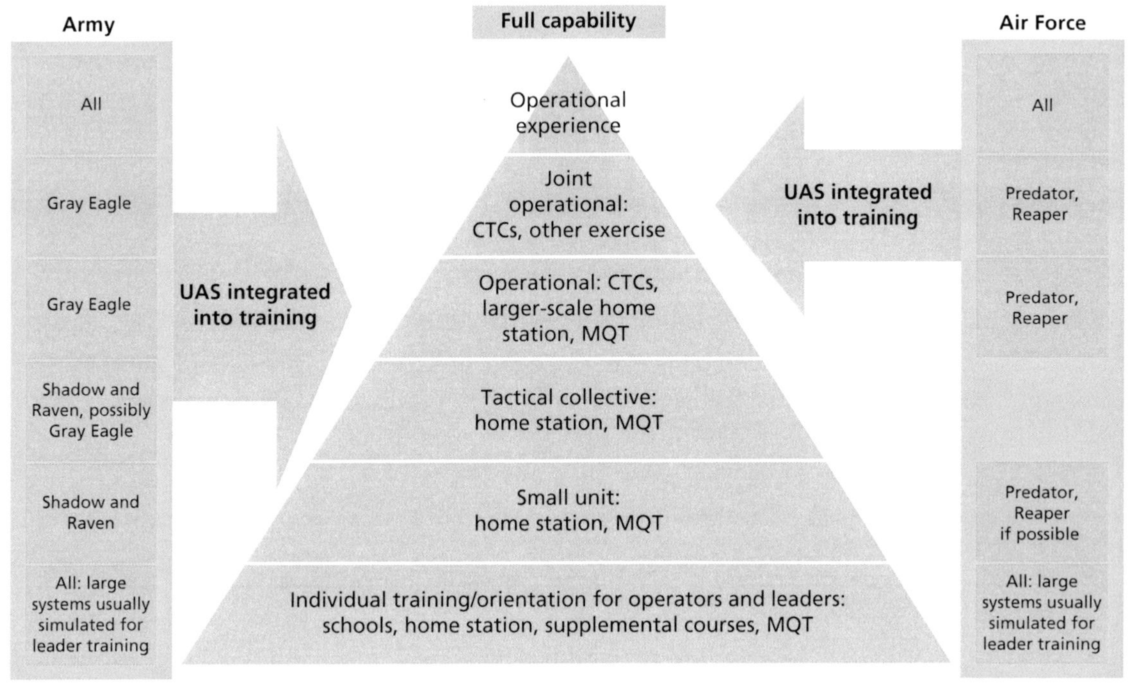

RAND *RR440-S.1*

operations involves Air Force units, which have gone through a similar process of progressive training. They join together in exercises, many of which are conducted at the large combat training centers (CTCs), to integrate close air support and Air Force UASs into the ground maneuver plan.

Assessment of the State of UAS Training in 2012: Service and Interoperability Training and the Role of Simulators

Service Training

Speaking generally, the services have been relatively effective at training the individual skills required to operate and maintain the UASs, although the services differ in their approaches. The Army, which tends to use enlisted personnel to staff its UAS units, has developed courses that award Military Occupational Specialties for the operators and maintainers for its Shadow and Gray Eagle systems. Training for the Raven UAS, which is about the size of a large model airplane and is used to support lower-echelon units, is a unit responsibility carried out through a "train-the-trainer" concept. The Marines use officers who are qualified aviators or aviation command and control officers as pilots, and enlisted personnel operate the payloads. The Air Force pilots are officers and are either rated pilots who graduated from pilot training or UAS-only pilots who graduated from a program designed specifically for flying UASs. Sensor operators for Reaper and Predator are either trained into a new career field or cross-trained from another Air Force Specialty Code. Those for Global Hawk come from the imagery analyst career field. Maintenance personnel come through the Air Force's maintenance education and training programs.

The more challenging part of the training comes at the integration level, where UAS personnel must meld their efforts with those of ground combat units. This occurs within and between services. Integration requires substantial practice under realistic conditions that do not occur frequently. Rotations at a CTC are often the only occasions, short of actual operations in theater, in which commanders face the challenge of integrating all battlefield operating systems at once, and the addition of the UASs necessarily complicates synchronization and integration tasks. Thus, it is not surprising that initial assessments of UASs during integrated training are, at best, mixed.

The key concerns of UAS units tend to be similar across services. For ground and Air Force units, the concerns are beddown and support facilities, airspace limitations, and simulators. Many of the beddown and support issues can be resolved with relatively minor construction. Some airspace issues, however, are not likely to improve. The Navy has airspace issues at specific installations but generally faces fewer restrictions because it can operate outside the 12-mile coastal limit with relative freedom.

Interoperability Training

To date, interoperability training at the National Training Center (NTC) and the Joint Readiness Training Center has been limited. The Government Accountability Office (GAO) has cited two significant challenges for improved interoperability training: (1) "establishing effective partnerships with program stakeholders through comprehensive communication and coordination and (2) developing joint training requirements that meet combatant commanders' needs," with particular emphasis on tactical-level training (GAO, 2005, p. 2).[1] To meet this challenge, the Department of Defense (DoD) established an implementation plan that gives the Office of the Under Secretary of Defense for Personnel and Readiness overall responsibility and the Deputy Under Secretary of Defense for Readiness executive agent responsibility for training transformation planning, programming, budgeting, and execution progress.[2]

While RAND did not assess interoperability training in detail, our discussions with the training community, including trainers at the NTC, suggest that such training opportunities have been limited because Air Force assets, particularly the Predator, have generally been unavailable due to pressing operational requirements in Iraq and Afghanistan. This same general observation was also reported by the GAO (2010b, p. 26). The problem of transiting Air Force UASs from Creech Air Force Base, Nevada, to the NTC at Fort Irwin, California, through Federal Aviation Administration–controlled airspace is often cited. Permanently stationing Air Force UAVs at the NTC, much as the Marine Corps has done by stationing Shadow UAS squadrons at the Marine Corps Air Ground Combat Center at Twentynine Palms, California, to support training there, would of course negate that problem. However, the experience of the Marine Corps suggests that the mere presence of UASs will not ensure

---

[1] GAO, 2005, p. 18, noted that,

> [i]n the past, joint training tasks were primarily focused at the command level and were identified through DOD's authoritative processes that built requirements by translating combatant commander inputs into training requirements. Training transformation has expanded joint training requirements to include those at the tactical level in addition to joint command-level training.

[2] See Director, Readiness and Training Policy and Programs, 2006.

that the forces are properly trained. The forces must be prepared for such training at their home stations. In addition, given the common MTTP, forces trained to operate with their own service UASs should find working with UASs from the other services less challenging.

Simulators

Congress has pressed for information on the role simulators might play in a UAS training strategy, seeking an "informed balance between live training and simulated training."[3] But the conditions that make a compelling case for using simulators in the training of fighter pilots are largely absent when it comes to UAS training:

1. UAS flying hours are much less expensive than flying hours for manned aircraft.
2. The kinds of operational activities that need more training emphasis, in particular air-ground coordination, are generally not activities for which simulators per se can substitute for live flying.
3. While simulators are an inherent part of the systems used for initial training of pilots and sensor operators,[4] they are not well suited (and not designed) for training operating forces on the full spectrum of UAS capabilities.
4. At best, simulators complement rather than substitute for live training. Given the current state of the fielding of UAS, first priority must be given to building the live flying infrastructure. Then, and only then, might funds be used to undertake the research and development that must precede any consideration of fielding the kind of interactive UAS air-ground simulators that could train both UAS crews and ground troops.

We concluded that, given current budget limitations and the importance of fully developing the opportunities for live training and the relatively low cost of such training, diverting funds to develop higher fidelity UAS-ground simulators would be unwise at this time. However, the Army's air-ground integration ranges provide a good example of judicious use of simulators to do what simulators do best—in this case, scoring target hits without using expensive live munitions and without blowing up fairly expensive targets. In the future, the services should reevaluate the need for virtual training based on a cost and effectiveness analysis (COEA) that includes developmental costs for simulators, savings based on the costs of flying UASs, and the critical availability of air space for training.

Implications and Recommendations

The Guiding Principle: "Train As We Fight"

DoD has been under some pressure from the Congress and such organizations as the General Accountability Office for the lack of central control and the lack of the development of joint DoD strategy for training. In reality, the services continue to develop UAS training initiatives primarily to meet their own unique requirements, and the opportunity for joint training is limited. Progress has, however, been made with the development of the MTTP and the recognition that interoperability training is a service responsibility. For now, DoD should encourage

[3] See, for example, Chairman, House Armed Services Committee, undated.

[4] See Shawn Johnson, UAS cost per flying hours, personal communication, April 10, 2013.

each service to solve its own UAS training problems, not constrain any one service under the guise of joint training. Given strong service programs, there will be ample opportunity to focus on joint training in the future, when the opportunity presents itself. Today, however, the services are still struggling with how to incorporate these new systems. The appropriate strategy for UAS training is to insist that the services train as they will fight. Accordingly, to *train as we fight* in the future, DoD training strategy must

- engender better appreciation of UAS capabilities throughout the chain of command
- address organizational, structural, and infrastructure and support issues
- enable "train as we fight" in collective unit training
- enable "train as we fight" in exercises.

Initiatives with Near-Term Payoff

To accomplish these goals, a number of initiatives should pay off in the near term, including

1. increasing exposure to the capabilities and limitations of UASs
2. harnessing the lessons-learned process to guide the development of service and joint doctrine and tactics, techniques, and procedures
3. addressing well-known but underresourced training infrastructure shortfalls.

Institutionalize Training for UAS Capabilities Over the Longer Term

Institutionalization will require more-general efforts, including

1. acculturation (Many end users do not yet have a well-developed appreciation of UAS capabilities.)
2. development of a cadre of UAS professionals with hands-on experience in all echelons of UAS operation
3. integration of UASs into service force structures
4. adaptation, including
   a. accounting for the evolution of the roles of UASs in the full range of military operations
   b. continuous resolution of UAS doctrinal issues.

Summary of Findings and Recommendations

The findings and recommendations are summarized below:

1. The proper strategy for DoD at this time is to encourage each service to solve its own UAS training problems, rather than to constrain any one of them under the guise of joint operations.
2. DoD should support current and future programs to develop ranges and beddown and support facilities similar to those in the Army's current programs.
3. In spite of the demands of deploying units and sustaining combat operations in Afghanistan, Army trainers indicate that Army units are better at maintaining qualification for Shadows than for Ravens. Nevertheless, not all units have been able to maintain high

Shadow qualification rates—some units arrive at the NTC with one-half or fewer of their operators qualified.

4. If joint training becomes a priority, the Air Force's current basing and beddown posture will become a problem. The Air Force could consider the location of Army hubs when choosing where to base its UAS fleet. A location that is near Army maneuver elements might make airspace access less of a restriction or at least make certificates of authorization more practical. Proximity might provide more opportunities for training UAS integration throughout the entire mission planning process.
5. Given current budget limitations and the importance of fully developing the opportunities for live training and the relatively low cost of such training, diverting funds to a research and development program to develop higher-fidelity simulators would seem to be unwise at this time.

Path to the Future

This report presents and discusses a multitude of means for setting up UAS training strategies for success, including supporting ongoing facilities and basing initiatives; expanding such facilities, where possible, to enable wider availability of UASs to support collective training; and increasing the use of UASs—the complete UAS package, including joint tactical air controllers and other coordinating elements—in collective training, both local and in larger exercises. We have noted the need to support continuing efforts to resolve airspace access issues, observing as we did so that there are ways to keep such restrictions from imposing serious limitations on much of the UAS training envisioned here. Similarly, we have discussed the potential for simulators to add value in training, now and perhaps more in the future. But we have also cautioned that simulators in their current state are not a good substitute for live use of UASs in collective training.

The path to the future for UAS strategies starts now, with support for ongoing initiatives that will continue the trends toward better training integration, thus improving the ability of end users to employ the multiple capabilities of UASs in their operations. The path continues, using that foundation, with longer-term efforts to add to acculturation of end users, professionalization of the UAS community, and integration of the two to harmonize the capabilities of UASs as key elements of overall force effectiveness.

Acknowledgments

The research presented here is the result of numerous field visits and discussions with members of service and department staffs in the Pentagon, commanders and operations staffs in operational arms of all the services, planners and training developers in service training commands, trainers and staffs at the various combat training centers, students in UAS operator schools, and UAS operators. So many shared information and data and providing thoughtful comments, advice, and insights that it is impossible to single out anyone without possibly slighting someone else. Moreover, the views expressed in this report are our own and should not be attributed to any individual or organization.

We would be remiss, however, if we did not recognize the support we received from Frank DiGiovanni, Director of Training Readiness and Strategy in the Office of the Under Secretary of Defense (Personnel and Readiness), and Joan Vandervort, our project coordinator in Mr. DiGiovanni's Office. Their commitment to developing an appropriate training strategy for UASs was the driving force behind this study.

Abbreviations

ABSAA	airborne sense-and-avoid
ACA	airspace control authority
ACO	airspace control order
ACTD	advanced concept technology demonstration
AECV	All-Environment Capable Variant
AFB	Air Force base
AFSC	Air Force specialty code
ALO	air liaison officer
AMB	air mission brief
AMPS	Aviation Mission Planning System
ARB	Air Reserve base
ATO	air tasking order
AVO	air vehicle operator
BAM	broad area maritime
BAMS-D	Broad Area Maritime Surveillance–Demonstrator
C2	command and control
CAOC	combined air operations center
CAP	combat air patrol
CAS	close air support
CCDR	combatant commander
COA	certificate of authorization
COCOM	combatant commander

CONUS	continental United States
CSG	carrier strike group
CTC	combat training center
DAS	Defense Acquisition System
DARPA	Defense Advanced Research Projects Agency
DDR&E	Director of Defense Research and Engineering
DoD	Department of Defense
DSB	Defense Science Board
FAA	Federal Aviation Administration
FAR	Federal Acquisition Regulation
FRS	fleet replacement squadron
FTU	formal training unit
FY	fiscal year
GAO	Government Accountability Office
GBSAA	ground-based sense-and-avoid
IQT	initial qualification training
ISR	intelligence, surveillance, and reconnaissance
JBLM	Joint Base Lewis-McChord
JFACC	joint force air component commander
JFC	joint force commander
JFCOM	Joint Forces Command
JIEDDO	Joint Improvised Explosive Device Defeat Organization
JRAC	Joint Rapid Acquisition Cell
JTAC	joint tactical air controller
LAA	limited acquisition authority
LCS	littoral combat ship
LRE	launch and recovery element
MCAGCC	Marine Corps Air Ground Combat Center
MCS	mission control system

MOS	military occupation specialties
MPO	mission payload operators
MTTP	mulltiservice tactics, techniques, and procedures
MQ-8B	Fire Scout
MQT	mission qualification training
NAS	National Airspace System
NCTE	Naval Continuous Training Environment
NDRI	RAND National Defense Research Institute
NTC	National Training Center
OEF	Operation Enduring Freedom
OIF	Operation Iraqi Freedom
PACOM	U.S. Pacific Command
PPBE	Planning, Programming, Budgeting, and Execution
REF	Rapid Equipping Force
RSTA	reconnaissance, surveillance, and target acquisition
RPA	remotely piloted aircraft
SOCOM	Special Operations Command
SOF	special operations forces
SOUTHCOM	U.S. Southern Command
SPINS	special instructions
SUAS	small unmanned aircraft system
TACP	tactical air control party
TEMF	tactical equipment maintenance facility
TF	task force
TRADOC	Training and Doctrine Command
UAS	unmanned aircraft system
UAV	unmanned aerial vehicle
UGS	Unmanned Ground School (Ft. Huachuca)
USD (AT&L)	Under Secretary of Defense (Acquisition, Technology, and Logistics)

USMC	U.S. Marine Corps
WSMR	White Sands Missile Range
YTC	Yakima Training Center

CHAPTER ONE

Introduction

UASs Play an Increasing Role on the Battlefield

During the last decade, in large part because of the dynamic nature of operations during Operation Enduring Freedom and Operation Iraqi Freedom, recognition of the importance of developing new capabilities to meet the ever-changing threats our military forces face has grown. As a result, the Department of Defense (DoD) has undertaken operational and technology demonstration projects to test new technologies and systems outside the traditional acquisition process. Many of these systems involve unmanned vehicles and some unmanned aircraft systems (UASs). In 2000, the DoD had fewer than 50 unmanned aircraft in its inventory; by 2012, it had more than 7,100, as shown in Table 1.1 (Gertler, 2012).[1] The size of the Department's investment in these systems today similarly dwarfs prewar levels. In 2000, the DoD spent $284 million on UASs, while the 2010 budget for UASs has grown to over $6.1 billion (GAO, 2010a, p. 1). Figure 1.1 shows the growth of UASs from 1988 to 2013. Figure 1.2 shows projected growth in UAS programs as of 2012.

UASs have two main advantages over manned aircraft: They eliminate the risk to a pilot's life, and they provide capabilities not subject to human limitations. They are also cheaper to procure and operate than manned aircraft. While they minimize risk to operational crews, they introduce new complications and hazards not associated with manned aircraft. While some think of UASs simply as a substitute for manned aircraft, UASs increasingly complement manned aircraft, providing new capabilities for the force to utilize. Combined UAS and helicopter operations are but one example (McLeary, 2012).

Originally, UASs were used to gather intelligence. During the Vietnam War, drones flew strategic reconnaissance missions over denied areas. The Israeli Air Force successfully used UASs during operations in Lebanon in 1982 and for many years led the world in developing UASs and tactics for employment. Subsequently, longer-endurance systems introduced the ability to maintain surveillance on distant and moving targets. More recently, UASs originally designed for reconnaissance have been modified to carry precision-guided weapons to attack ground targets, greatly expanding the role such systems play on the modern battlefield. This also has expanded and complicated the training of UAS crews as full participants in the joint and combined arms battle.

Successful operational tests and demonstrations of the expanded range of UAS capabilities have led to rapid fielding of new systems, often placing unanticipated demands on logistics

[1] See Appendix A for descriptions of the major systems.

Table 1.1
DoD UAS Platforms as of 2011

Name	Vehicles	Ground Control Stations	Employing Service(s)	Capability/Mission
RQ-4A Global Hawk/ BAMS-D Block 10	9	3	USAF Navy	ISR Maritime domain awareness (Navy)
RQ-4B Global Hawk Block 20/30	15	3	USAF	ISR
RQ-4B Global Hawk Block 40	1	1	USAF	ISR Battle management command and control
MQ-9 Reaper	54	61	USAF	ISR RSTA EW Precision strike Force protection
MQ-1A/B Predator	161	61	USAF	ISR RSTA Precision strike Force protection (MQ-1C Only-C3/LG)
MQ-1 Warrior/MQ-1C Gray Eagle	26	24	Army	ISR RSTA Precision strike /force protection (MQ-1C Only-C3/LG)
UCAS-D	2	0	Navy	Demonstration Only
MQ-8B Fire Scout vertical takeoff and landing tactical UAV	9	7	Navy	ISR RSTA Antisubmarine warfare Antisurface warfare Mine warfare Organic mine countermeasures
MQ-5 Hunter	25	16	Army	ISR RSTA Battle damage assessment
RQ-7 Shadow	364	262	Army USMC SOCOM	ISR RSTA Battle damage assessment
A160T Hummingbird	8	3	SOCOM DARPA Army	Demonstration
Small tactical UASs	0	0	Navy USMC	ISR Explosive ordnance disposal Force protection
ScanEagle	122	39	Navy SOCOM	ISR RSTA Force protection
RQ-11 Raven	5,346	3,291	Army Navy SOCOM	ISR RSTA
Wasp	916	323	USMC SOCOM	ISR RSTA
SUAS AECV Puma	39	26	SOCOM	ISR RSTA
Gasoline-powered micro air vehicle (gMAV) T-Hawk	377	194	Army (gMAV) Navy (T-Hawk)	ISR RSTA Explosive ordnance disposal

SOURCE: Gertler, 2012, p. 8.

Figure 1.1
UAS Budgets from 1988 to 2013

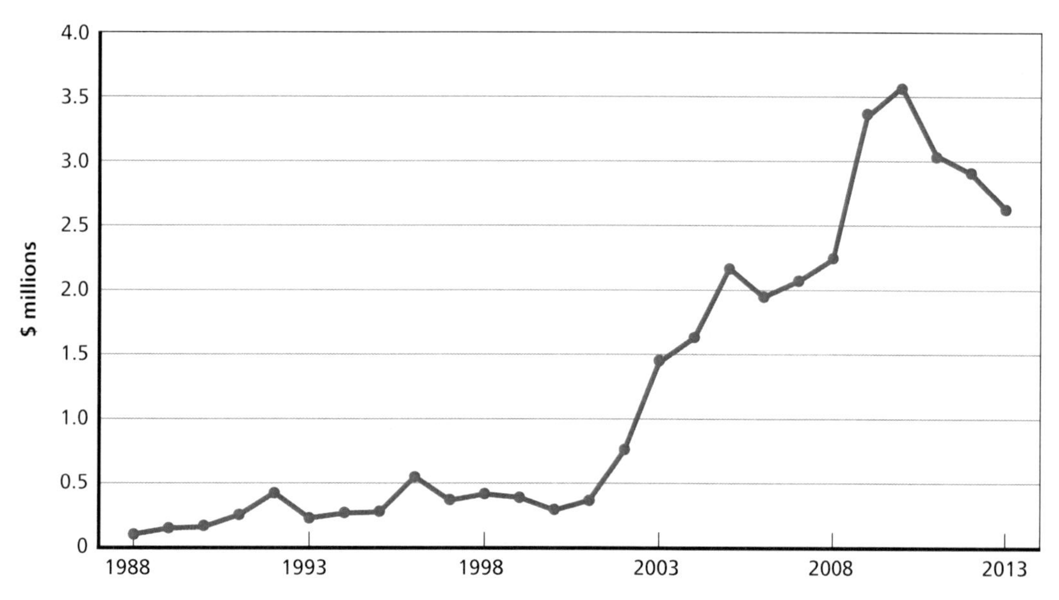

SOURCE: Gertler, 2012, p. 14.
RAND *RR440-1.1*

Figure 1.2
DoD Inventories for Medium and Large Unmanned Aircraft

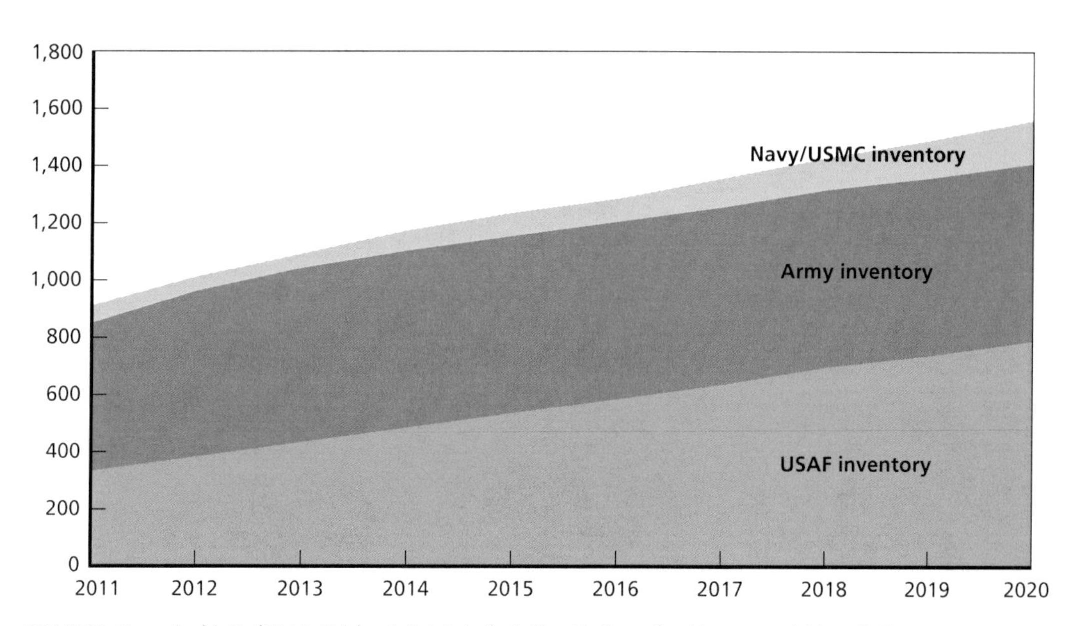

SOURCE: Kempinski, B. (2011, Tables 1-2,1-3,1-4). Policy Options for Unmanned Aircraft Systems, Congressional Budget Office.
RAND *RR440-1.2*

and training systems and on field commanders to employ the new systems effectively. UASs must now be integrated into the training programs of the services.

Building a responsive, effective, and efficient UAS training program is a challenge in a time of reduced budgets. Any new program must be based on a review of existing training capabilities and new investments across the services.

Integrating UASs into the core operations of the DoD presents the kinds of special problems that are common when new technologies appear and clash with existing ways of doing business. Such clashes happen frequently enough that there is a term for such innovations: disruptive technologies. The rapid introduction of UASs into operations in operations Enduring Freedom and Iraqi Freedom and the attendant force structure growth have presented just such challenges (as discussed in Chapter Two). The resulting disruption has contributed to lags in the ability of joint doctrine, tactics, techniques, and procedures (TTP) to keep pace and in the ability of joint training[2] and service-specific managers to develop and execute training programs.[3]

Focus and Approach of this Research

The Deputy Director, Readiness and Training Policy and Programs in the Office of the Under Secretary of Defense for Personnel and Readiness asked RAND to assess the adequacy of UAS training to support current and future requirements. Proposals to resolve service and joint UAS training issues must be informed by a clear understanding of current problems, opportunities for correction, and associated costs and benefits of the corrections. In addition, the current extraordinary pressures on the overall federal budget, including the defense "top line," and the ongoing review of U.S. strategy will likely have implications for force structure and basing, resulting in a number of yet-unknown factors that could profoundly affect the future of UAS programs. These uncertainties notwithstanding, a number of issues must be addressed: (1) a general concept for UAS training; (2) an appropriate framework for addressing UAS training requirements, including the use of simulators; and (3) the airspace requirements necessary for UAS training.

The research reported on here covers UASs in the Army, Navy, Air Force, and Marine Corps as fielded and plans as they existed in 2012. Because of the increasing use of these unmanned systems to support to ground operations, the RAND effort has more closely examined the interface between UASs and operating elements of ground forces. To this end, a team from RAND carried out extensive field visits to understand the current ability of the services to conduct (1) service-specific training and (2) joint training at both home station and joint train-

---

[2] Department of Defense Directive 1322.18, January 13, 2009, defines joint training as "Training, including mission rehearsals, of individuals, units, and staffs using joint doctrine or tactics, techniques, and procedures to prepare joint forces or joint staffs to respond to strategic, operational, or tactical requirements that the Combatant Commanders (CCDRs) consider necessary to execute their assigned or anticipated missions."

[3] Chairman of the Joint Chiefs of Staff Instruction 3500.01G, March 15, 2012, p. C-2, defines service training as follows:

> Service Training: Service Active Component (AC) and RC training (including USSOCOM) is based on joint and Service policy and doctrine. Service training includes basic, technical, operational, and interoperability training to both individuals and units in response to operational requirements identified by the CCDRs to execute their assigned missions.
>
> Joint Training. Training, including mission rehearsals, of individuals, staffs, and units, using joint doctrine and tactics, techniques, and procedures, to prepare joint forces or joint staffs to respond to strategic, operational, or tactical requirements considered necessary by the CCDRs to execute their assigned or anticipated missions.

ing facilities. Members of the team interviewed officials involved with UAS training operations at the following Army, Navy, Air Force, and Marine installations and units:

- Beale Air Force Base (AFB), California
  - Global Hawk operations and Distributed Common Ground Station 2
- March Air Reserve Base (ARB), California
  - Air Force National Guard Ground Control Station school
- Holloman AFB, New Mexico
  - (Air Force Predator and Reaper training)
- Fort Irwin, California
  - (Air Force training support unit and selected individuals from the operations group)
- Creech AFB, Nevada
  - (the Joint Unmanned Aircraft Systems Center of Excellence and the operational UAS wing)
- Fort Rucker, Alabama
  - Training and Doctrine Command (TRADOC) capability manager for UASs
- Fort Huachuca, Arizona
  - TRADOC Capability Manager for Intelligence Sensors and the UAS Training Battalion
- Fort Benning, Georgia
  - Maneuver Center of Excellence
- Fort Campbell, Kentucky
  - 101st Airborne Division (Air Assault)
- Marine Corps Base Twentynine Palms, California
  - Marine Corps Air Ground Combat Center (MCAGCC) and two of three active U.S. Marine Corps (USMC) UAS squadrons
- Naval Air Station Patuxent River, Maryland
  - Naval Air Systems Command (Navy and Marine Corps)
- Fort Eustis, Virginia
  - TRADOC Capability Manager for Live Fire Training, Army Capabilities Integration Center, and TRADOC G-3
- North Island Naval Base, San Diego, California
  - Commander, Naval Air Forces.

In addition, the RAND team had extensive discussions with the strategy and planning staffs at U.S. Pacific Command (PACOM) and U.S. Navy Pacific Command concerning rapid acquisition of new technologies, such as the UASs reported on here. In that process, the RAND team gained important insights into the challenges the services face, first in identifying new or disruptive technologies and then in using operational demonstrations or experiments to understand and develop the changes to operational concepts that these technologies would drive.

During these visits, the RAND team talked with operators and support personnel; those who train operators and associated members of the team; those who train and observe the forces that employ UASs in their operations; staff elements concerned with the planning and resourcing of training, both for UAS operators and for the force more generally; and those concerned with the ongoing development of doctrine and concepts of operations for UASs, including their integration into operations.

Plan for the Report

Chapter Two starts the report with a discussion of disruptive technologies, rapid acquisition processes, and the challenges the DoD faces in integrating UASs into the operations of the military departments. Chapter Three presents a general concept and a framework for training focused on the integration of UASs into military operations. Using the framework, the assessment in Chapter Four considers the current ability of the services to conduct service-specific training at both home station and joint training facilities. This includes a compilation of observations and insights largely gained during our visits to the bases listed above, with our interpretation of their implications for the specification and resourcing of training strategies. Finally, Chapter Five summarizes our research and looks toward the evolving training strategy for DoD in the future. The appendixes present additional material, including a description of ongoing systems, a further discussion of the Defense Acquisition System (DAS), UAS descriptions, and infrastructure considerations and the results of a RAND assessment of training infrastructure.

CHAPTER TWO

The Case for and Implications of Unmanned Aircraft Systems Being a "Disruptive Technology"

Over decades, DoD has developed a structured and deliberate approach for the way it acquires new systems, which is formalized in the DAS and the 5000 series of DoD instructions. However, the exigencies of recent military operations have led DoD to develop new technology and acquire and field new systems without going through the steps the defense acquisition process generally requires. While near-term requirements drove such acquisition decisions, these decisions have long-term consequences. Rapid acquisition without a fully developed plan for how the new systems will be sustained over time may well diminish operational effectiveness. A potent example of such an effect is the extent to which the rapid and large-scale introduction of UASs has outpaced investments in the training infrastructure necessary to support the associated operation and maintenance. In the future, personnel will not be able to use or maintain the new systems properly. The Defense Science Board (DSB) made this point forcefully in 2003: "(1) military proficiency is as dependent on the warriors who operate weapon systems as it is on the weapon system technology, and (2) a superb way to waste personnel or system acquisition money is to ignore training" (DSB, 2003, p. 2). Despite the inefficiency of poorly coordinated investments, the DSB found that "systems continue to be created and then fielded with little consideration for the costs that must be incurred during the life cycle to train the weapon's users" (DSB, 2003, p. 44). As it turns out, the difficulty of incorporating new technologies into existing operations and processes is not unusual. It happens frequently enough that there is a term for such innovations: *disruptive technologies*.

The Disruptive Technology Experience in the Business Community

Clayton Christensen introduced the term *disruptive technology* in 1995. He distinguished disruptive technology from what he termed *sustaining technology*. In his nomenclature, a sustaining technology improves the performance of an existing system and does not require significant structural adjustments to processes, organizations, or operational paradigms. Disruptive technologies, however, change the way a business operates (Govindarajan, Kapalle, and Daneels, 2011).

Since the concept was first introduced, a number of studies have focused on ways that enable an organization to incorporate disruptive technologies. Christensen found that many organizations fail to meet the challenges disruptive technologies pose because they focus only on costs and not on how the new technologies change business processes and create new values for customers (Christensen and Overdorf, 2000). As Christensen used the term, *pro-*

cesses include the coordination, communication, decisionmaking, and interaction patterns that transform resources into a product. Under normal conditions, processes are stable. They are designed not to be easy to change, and this inertia can impede the adoption and exploitation of emerging disruptive technology (Looy, Martens, and Debackere, 2005). Tellis makes the point that, given limited resources, investments in unfamiliar technologies with unproven returns are opposed because they are seen as "cannibalizing" both resources and customers from established products (Tellis, 2006). Because it is often difficult to introduce such new technologies into existing business units, much of the literature discusses how "spinout" organizations were created to nurture these new technologies (e.g., Henderson and Clark, 1990). For example, in a study of several game-changing technological developments, te Kulve and Smit, 2010, found that successful new product lines usually had a "champion" with "dedicated change strategies" to further the introduction of disruptive technologies.

Unmanned Aircraft Systems: A Case of Disruptive Technology for the Department of Defense

Although DoD operates in a different environment with different incentive structures from those of the business community, it is not immune to the difficulties disruptive technologies have introduced in the private sector.[1] From DoD's perspective, a technology is disruptive when it has the potential to alter the scope and effectiveness of military operations dramatically and is applied in sufficiently large numbers that it can no longer be supported in the same manner as during its initial development (i.e., the military must incorporate its support into its standard logistics and training systems). The UAS qualifies as a disruptive technology on multiple counts. First, as Christensen notes, the advent of the UAS changed the way the U.S. military operated, that is, its "business processes" changed. Second, UASs have broadened the scope of military operations substantially and greatly enhanced the effectiveness of both intelligence processes and strike operations. Finally, they have arrived in sufficient numbers that they now need to be integrated into the standard processes.

As the U.S. went to war in Afghanistan and Iraq, concern grew about the lack of speed and responsiveness in the traditional process for getting new systems into the hands of warfighters.[2] Critics complained about bureaucratic processes that make the traditional system too slow and too risk averse to acquire and field effective solutions rapidly.[3] As late in the conflicts as 2008, Secretary of Defense Robert Gates expressed frustration with the defense bureaucracy, arguing "we must not be so preoccupied with preparing for future conventional and

[1] There are numerous examples of disruptive technologies and how they displaced formerly dominant technologies in both the private sector and the military: Just a few examples are commercial airlines displacing passenger trains, internal-combustion-engine vehicles replacing horse-drawn vehicles, aircraft carriers replacing battleships as the Navy's capital ship, and tanks replacing cavalry.

[2] Appendix B describes DoD's traditional acquisition system.

[3] For example, Lt. General Robert P. Lennox, the Army's Deputy Chief of Staff for Programs and Resources, explained, "This is one of our most common topics of conversation: How do we rapidly field capabilities when technology advances so quickly?" (quoted in Erwin, 2010).

strategic conflicts that we neglect to provide both short-term and long-term all the capabilities necessary to fight and win conflicts such as we are in today."[4]

A Challenge to the Traditional Acquisition System

The virtues of the deliberate and structured traditional acquisition system must be weighed against the fact that the system can take years to field new capabilities. From requirements generation through the DAS and the Planning, Programming, Budgeting, and Execution (PPBE) process, the traditional acquisition system generally takes several years to meet materiel warfighter needs; some systems can take decades (GAO, 2010b). While this long time frame makes some sense for large and complex programs of record, it poses a significant challenge when required capabilities are needed more quickly or are focused more narrowly. This lack of responsiveness has led some observers to argue that the traditional system is not optimized for delivering the full spectrum of desired capabilities. While the system should, in theory, be adaptable for both small and large acquisition programs, it works best in practice for major defense acquisition programs. However, much recent acquisition activity has not been in the form of large programs of record that entail the institutional backing necessary for proper sustainment strategies. The fast rate and large volume demanded for unfamiliar technologies, in particular, have strained the system and pressured traditional acquisition relationships to change.

Rapid Acquisition Programs Met the Challenge

To meet an urgent demand to incorporate new technologies, a series of acquisition processes was created outside the traditional acquisition process and was unencumbered by the prior restrictions designed to mitigate cost, capability, and schedule risk. Unfortunately, while these "workarounds" fostered the fielding of numerous systems, they generally neglected to provide for the institutional sustainment—training and logistical support—needed to facilitate introducing these systems into the force on any basis other than as technology demonstrations. Figure 2.1 captures the proliferation of a selection of these rapid acquisition processes. Below, we discuss several of the more important ones.

Advanced Concept Technology Demonstration

In 1994, in response to recommendations from the Packard Commission and the DSB, DoD introduced the advanced concept technology demonstration (ACTD) (Drezner, Sommer, and Leonard, 1999, p. 23). The Global Hawk UAS, which has found broad applicability in the wars in Iraq and Afghanistan, began as an ACTD. This approach to technological development was intended to improve speed and promote innovation by limiting acquisition hurdles. A system designated as an ACTD could use streamlined management and had reduced oversight requirements (Drezner, Sommer, and Leonard, 1999, p. xiv). ACTDs were intended to bridge the innovation-adoption gap by existing somewhere beyond demonstration of technical feasibility without quite yet becoming major defense acquisition programs. Toward the same end of nudging innovation and eliminating procedural hurdles, in 1994 the National Defense

---

[4] Gates' distinction reflects what is often a difficult balance between short- and long-term requirements and planning horizons. This creates an institutional tension that has, for example, made the military departments resistant to prioritizing the acquisition of unmanned technology that had near-term applicability but that was not intended to be part of a longer-term vision. See Gates, 2008.

Figure 2.1
Select Rapid Acquisition Processes, Approaches, Authorities, and Funding Mechanisms

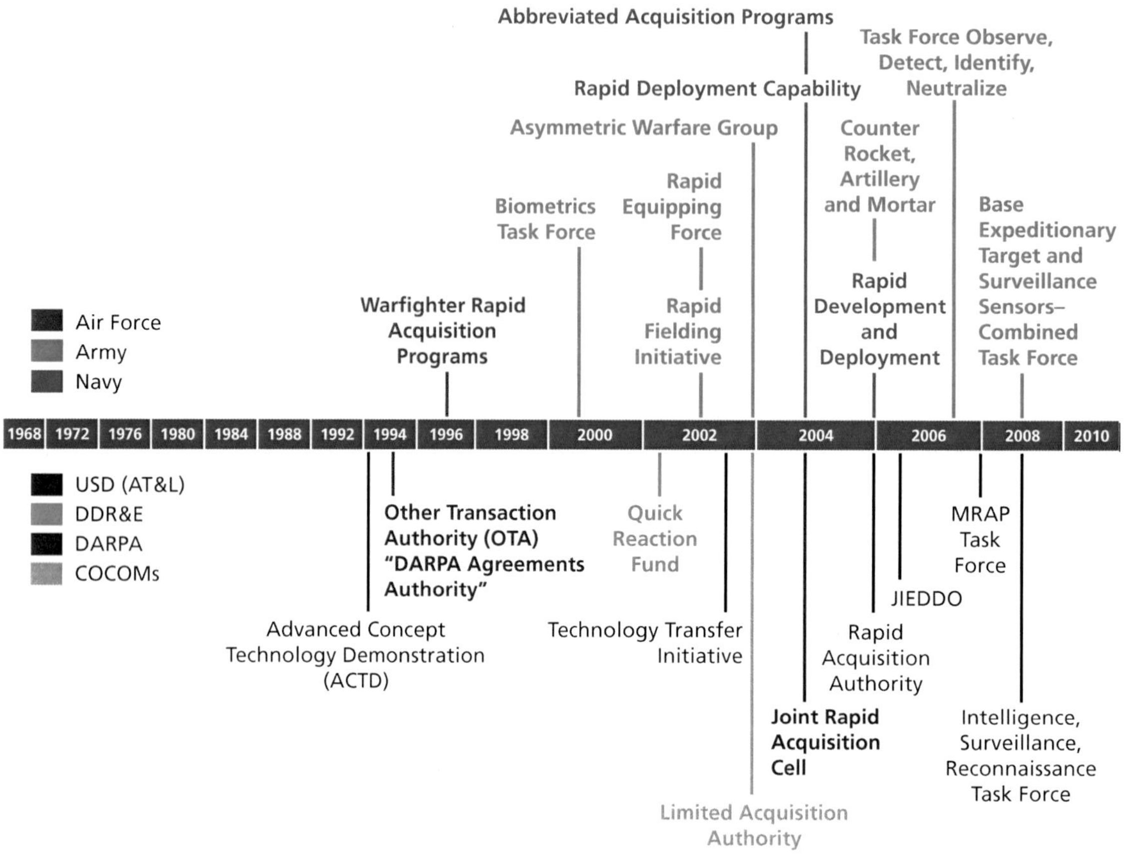

Abbreviated Acquisition Programs
Task Force Observe, Detect, Identify, Neutralize
Rapid Deployment Capability
Asymmetric Warfare Group
Counter Rocket, Artillery and Mortar
Rapid Equipping Force
Biometrics Task Force
Base Expeditionary Target and Surveillance Sensors–Combined Task Force
Rapid Development and Deployment
Warfighter Rapid Acquisition Programs
Rapid Fielding Initiative
Air Force
Army
Navy
1968 1972 1976 1980 1984 1988 1992 1994 1996 1998 2000 2002 2004 2006 2008 2010
USD (AT&L)
DDR&E
DARPA
COCOMs
Other Transaction Authority (OTA) "DARPA Agreements Authority"
Quick Reaction Fund
MRAP Task Force
JIEDDO
Advanced Concept Technology Demonstration (ACTD)
Technology Transfer Initiative
Rapid Acquisition Authority
Joint Rapid Acquisition Cell
Intelligence, Surveillance, Reconnaissance Task Force
Limited Acquisition Authority

SOURCE: Derived from Anderson (1992), AUSA (2003), DSB (2009b), GAO (1998), GAO (2010b), Sullivan (2009a), Thirtle (1997).
RAND *RR440-2.1*

Authorization Act introduced an authority that allowed the Defense Advanced Research Projects Agency (DARPA), long seen as a mechanism for moving innovative ideas toward materiel solutions, more flexibility in contracting; facilitated use of commercial practices; and waived certain Federal Acquisition Regulations (FARs) and other laws (Drezner, Sommer, and Leonard, 1999, p. 27).

Limited Acquisition Authority

In time of war, new requirements arise that the long and deliberate acquisition system is ill suited to meet. As a result, both Congress and DoD's civilian leadership put in place ad hoc arrangements to seed the development and fielding of new systems. In fiscal year (FY) 2004, Congress granted the Secretary of Defense the authority to delegate a limited acquisition authority (LAA) to Joint Forces Command (JFCOM) to speed the research and development and limited fielding of new systems (GAO, 2007a). However, LAA is an authority, rather than a program, and does not translate into budgeted funds.[5] In the first three years of the authority,

[5] JFCOM had to reallocate funding from its own budget or gain funds from another DoD organization. Over one-half the programs JFCOM supported came from its own budget. In addition, after the systems were acquired, funds were not

JFCOM used the LAA to support six projects. After that, LAA activities slowed significantly, a trend GAO suggests may be linked to lack of access to funding (GAO, 2007a, p. 3).

Joint Rapid Acquisition Cell

The Joint Rapid Acquisition Cell (JRAC), also established in 2004 by the Deputy Secretary of Defense and organized in the Office of the Under Secretary of Defense for Acquisition, Technology, and Logistics, constituted another rapid acquisition program. The intent of this cell was to meet new requirements that the combatant commanders (COCOMs) identified as operationally critical; the COCOMs' prominent role in JRAC has led some to tag it the "COCOM's acquisition process" (Middleton, 2006, p. 19; see GAO, 2010b, p. 1). During its initial three years, JRAC supported 24 programs, at a cost of $335.5 million.[6] These funds came primarily from war supplemental appropriations and were not part of the military departments' programs or budgets. A 2009 DSB report found that this meant that "these programs continue to lack serious institutional commitments; very little is being built into the service or other DOD budgets for these programs" (DSB, 2009, p. 6). While JRAC required a life-cycle plan as part of the acquisition process, the DSB found this planning inadequate. Training relied more "on learning on the job with little emphasis on support, training, and sustainment." (DSB, 2009, p. 6). The typical institutional support in the services that is needed to fund, for example, investments in the training infrastructure does not exist for every rapidly acquired system.

Rapid Equipping Force

In 2002, the Army established the Rapid Equipping Force (REF), whose mission was to equip operational commanders with technological solutions to urgent war needs in no more than six months (DSB, 2009, p. 12). In 2005, the REF was made permanent (Kennedy, 2006, pp. 43–45). The organization began with a staff of 14 personnel, which expanded to 150 by 2007 (Dietrich, 2007, p. 5). In 2005 alone, it purchased more than 20,000 items, including robots, surveillance systems, digital translators, and weapon accessories (Dietrich, 2007, p. 6). The REF supported the rapid acquisition and fielding of a widely adopted technology called PackBot, which became a vital tool in the identification and defusing of deadly improvised explosive devices. The REF also supported other unmanned and robotic technology, including the development of the battery-powered, hand-launched, and camera-equipped Tactical Mini Unmanned Aerial Vehicle (UAV) in 2005. The REF project leader described that UAV as a technology that allowed soldiers to "go into situations knowing what's in front of them" (Ainsworth, 2005). REF purchased the mini-UAV commercially and, after some software modifications, quickly had it ready for theater operations (Miles, 2005). Yet, insufficient consideration for back-end support reportedly plagued early REF efforts, and steps have been taken to ameliorate the difficulties. COL Gregory Tubbs, REF director, conceded that the early ad hoc "wild west" days of REF logistical support were a challenge (Kennedy, 2004).

---

readily available for long-term sustainment. A 2005 report from the Center for Strategic and International Studies called for providing rapid acquisition approaches with more-secure funding: "Urgent requirements will be met much faster if they can be resourced without taking funds from existing programs" (Murdock and Flournoy, 2005, p. 98).

[6] The process begins when a COCOM identifies and validates an urgent operational needs statement (or a joint urgent operational needs statement for joint requirements), defined as "urgent, combatant commander-prioritized operational needs that, if left unfilled, could result in loss of life and/or prevent the successful completion of a near term military mission." *Immediate warfighter needs* were the urgent operational needs statements certified as requiring a solution in 120 days or less. For a given system, JRAC is authorized to expend $365 million in research, testing, development, and evaluation funds and up to $2.19 billion in procurement. See GAO, 2007a, p. 12.

Intelligence, Surveillance, and Reconnaissance Task Force

The high value placed on UASs, specifically in new intelligence, surveillance, and reconnaissance roles, can be seen in the robustly funded Intelligence, Surveillance, and Reconnaissance (ISR) Task Force, which supported the fielding of numerous unmanned systems. In spring 2008, Secretary Gates established this task force to expedite the fielding of ISR assets to combat areas (Sherman, 2008). The task force originally had a 120-day charter; two years after the task force's founding, the secretary announced that it would become a permanent part of the Office of the Under Secretary of Defense for Intelligence (Bennett, 2010). Its mission was to address unmet ISR requirements and to rapidly acquire and field capabilities by coordinating activities and pursuing innovative solutions (Loxterkamp, 2010). The ISR Task Force was to recommend ways to maximize the availability of systems in the inventory and to boost acquisition of additional systems (Best, 2010, p. 17). In FY 2008 alone, Congress approved the reprogramming of $1.3 billion based on ISR-TF recommendations (GAO, 2008b). Among the systems the task force supported were the Navy's unmanned MQ-8 Fire Scout and the fielding and sustainment of 50 unmanned Predators.

Informal Approaches to Rapid Acquisition

The formal approaches highlighted above support rapid acquisition in a variety of ways. Some consist of an acquisition authority but no money (LAA); some consist of funding (REF); some facilitate stronger requirements inputs (JRAC); and some, like Special Operations Command (SOCOM), have everything. Yet importantly, some rapid acquisition has taken place outside even these ad hoc institutional arrangements. For example, in one case, the personal relationships enabled rapid acquisition of command and control (C2) systems. In another, ADM James Stavridis instituted a "culture of innovative thought" into U.S. Southern Command (SOUTHCOM) by developing his own in-house technical capabilities through a process he called Linking Plans to Resources, an approach aiming to associate capabilities with specific solutions (Hicks, 2008, p. 33). He chartered the Joint Innovation and Experimentation Directorate, tasking it with improving how the command trains, fights, and does business (Stavridis, 2010, p. 177). This small innovation staff was to "research, explore, and test emerging technologies available commercially or through Federal research centers" (Stavridis, 2010, p. 89). The directorate was to take the lead in identifying new and creative ways of meeting the command missions, and was given the primary responsibility for "developing validated solutions into an initial operational capability," materiel or nonmateriel (Stavridis, 2010, pp. 178–179). These innovations, looking for better ways to link requirements to resources and developing in-house capabilities for acquiring solutions, reflected new acquisition roles.

PACOM's version was the Plans to Resources to Outcomes Process, which helped in the development of the command's integrated priority list. PACOM also developed its own in-house means of meeting requirements (Murdock and Flournoy, 2005, p. 40). PACOM's Joint Innovation and Experimentation Division uses the integrated priority list to develop rapid, innovative solutions for filling the gap and injecting them into PACOM exercises. Similarly, in 2009, U.S. European Command appointed a special assistant for innovation and technology to field equipment for the war in Afghanistan. Navy Captain Jay Chestnut, special assistant in charge of the project, explained that European Command "knows that 'big acquisition' [the traditional system] is trying to do the right thing but sometimes you need someone working on the side, looking innovatively" (Francis, 2009).

In addition to the development of in-house approaches, informal rapid acquisition has also involved forging relationships between the COCOMs and technical expertise outside the military departments. As described by Stavridis, SOUTHCOM worked to both build in-house staff and forge relationships with a "technological base" external to the organization. He highlights, in particular, the partnership with DARPA, in which DARPA pursues "exploration and technology where risk and payoff are both very high, and where success may provide dramatic advances" to the SOUTHCOM mission (Stavridis, 2010, p. 89). Included in Admiral Stavridis' list of outcomes of this productive symbiotic relationship with DARPA are unmanned aerial craft and unmanned surface vessels.

The Use of Fast Track Authorities Can Cause Problems

For acquisitions to be "rapid," certain aspects of the traditional acquisition process have been left out. In many cases, it is the training and support planning that have been omitted. In 2008, the GAO reported that the

> rapid fielding of new systems and the considerable expansion of existing Air Force and Army programs has [sic] posed challenges for military planners to fully account for UAS support elements, such as developing comprehensive plans that account for personnel and facilities needed to operate and sustain programs." (GAO, 2010a, p. 37)

Because many of these programs do not pass through the system development and demonstration phase or a logistics supportability analysis, the data needed to generate training and support planning and ultimately guide investment decisions are unavailable. In 2010, the GAO found that the Air Force had not developed a servicewide plan that identified the number of personnel to be trained, the specific training required, and the resources necessary to establish a dedicated UAS training pipeline (GAO, 2010a). The Air Force has reportedly been struggling to staff significant increases in unmanned systems (Baldor, 2010). Going from a handful of drones in 2007 to 45 by 2010, with plans to operate 50 by 2011 and 65 by 2013, had created significant resource challenges for the department. These resource challenges became heavily pronounced with the "surge" in UAS requirements for operations in Libya and Afghanistan beginning in March and extending into summer 2011. The Air Force, to fulfill combat air patrol (CAP) requirements, stood down a portion of its formal training structure to form three CAPs (U.S. Air Force, 2011). As the Air Force's Deputy Chief of Staff for Operations, Plans, and Requirements explained in 2010, the "number one manning problem in our Air Force is manning our unmanned platforms" (Baldor, 2010).

The GAO similarly found that the Army's personnel authorizations were insufficient to support UAS operations. The Army has determined on at least three separate occasions since 2006 that Shadow UAS platoons did not have adequate personnel to support the near-term and projected pace of operations (GAO, 2010a). Officials from seven Army Shadow platoons in the United States and Iraq told the GAO that approved personnel levels for these platoons did not provide an adequate number of vehicle operators and maintenance soldiers to support operations (GAO, 2010a). Army officials told the GAO that currently approved personnel levels for the Shadow platoons were based on planning factors that assumed that the Shadow would operate for 12 hours per day with the ability to extend operations to up to 16 hours for a limited time (GAO, 2010a). However, personnel with these platoons told the GAO that UASs in Iraq routinely operated 24 hours per day for extended periods (GAO, 2010a). Army officials also reported that combat brigades and divisions require additional personnel to provide UAS expertise to assist commanders in making effective use of new technological systems (GAO, 2010a).

UASs exemplify the challenges of integrating training and sustainment considerations outside the strictures of the formal acquisition process. Ad hoc rapid acquisition processes were able to field UASs rapidly, but because of the nature of these processes, the fielded UASs lacked sustainable institutional support and funding needed for future investments in the training infrastructure. Frequently, the services try to sustain and support these programs by relying on contractors, just as they did during system development.[7] Of course, employing contractors to provide training is not in itself a problem. But it does reflect the extent to which training has been more about expediency than about effective institutionalization, particularly with regard to the imperative to train as we fight and the need to fully appreciate the need to integrate UAS training across the full spectrum of combat units.

---

[7] For example, the Air Force continues to rely on contractors to perform a considerable portion of UAS maintenance. For example, contractors perform approximately 75 percent of organization-level maintenance requirements for the Air Combat Command's Predator and Reaper UASs. See GAO, 2010a.

CHAPTER THREE

Training Concept and Framework for Unmanned Aircraft Systems

We will begin by discussing the general ideas that should guide UAS training—the training *concept*. Then, we will consider how the parts identified in the concept fit together into a conceptual structure for that training—a training *framework* designed to enhance combat power and other operational capabilities. Note that the term *UAS* can be misleading. In fact, there is not a single UAS but rather a family of aircraft that share the common feature that they fly without a pilot on board.[1] These aircraft have very different personnel and training requirements. The vast majority of UASs support ground operations, and that will be our primary focus. As noted in Table 1.1, ground forces—the Army and Marine Corps—own many of these systems themselves, but the Air Force owns some of them and often flies them in support of ground operations. UASs are sometimes flown in support of conventional ground operations and sometimes in support of special operations. Our main focus is on support of conventional forces. UASs are now also being developed that will support maritime forces, and we also consider them.

Training Transformation, Joint Doctrine, Joint Operations, and UASs

Large military operations today are usually joint operations requiring and emphasizing the interdependence of the services. It follows naturally that joint doctrine, joint concepts of operation, and joint TTP would provide focus for training. Accordingly, the DoD's Training Transformation Implementation Plan requires that in order to "improve joint force readiness" there be a "unity of effort in training across Services, agencies and organizations."[2] That said, joint training guidance and publications generally provide guidance direction on the coordination and collaboration among elements of different services to optimize the joint employment of their various capabilities, and not on the employment of any one, specific system or even family of systems, such as UASs.

Clearly, training does not take place for the sake of training; it should be driven by doctrine. Thus, in line with the general focus described above, JP 3-0 (2001) "establishes the framework for our forces' ability to fight as a joint team," but nowhere does the document discuss the management of UASs or any other specific system. Moreover, nowhere in any other joint publication is there a grand doctrine for the employment or management of UASs on

[1] In 2014, the Air Force republished its vision for its system, replacing the term *unmanned aircraft systems* with the term *remotely piloted aircraft* (RPA) (U.S. Air Force, 2014, p. iii).

[2] See Director, Readiness and Training Policy and Programs, 2006, p. 5.

the battlefield. However, the services have agreed to a set of TTP to be incorporated into their respective training programs. They describe these as a "multiservice" rather than a joint TTP, thereby avoiding doctrinal issues on how to deploy and manage UAS even as the services support joint operations.

Air Force Maj David Buchanan explored the difficulty of developing a joint doctrine for UASs in a 2010 Naval War College paper, summing up the difference this way: The Air Force believes in centralized control, and the Army believes in decentralized operations. He noted that the

> Air Force's proposal to assume executive agency for all medium- and high-altitude UASs in March 2007 was an attempt to establish unity of command and increase UAS efficiencies. ... The Army's position ... was that a single-service approach to UAS employment would infringe on the effectiveness of UASs in combat.
>
> The Air Force's centralized approach may result in a more efficient allocation of limited UAS assets, but that efficiency comes at the cost of combat effectiveness. The Army's answer to regain effectiveness has been to decentralize command and control for its unmanned aircraft systems. Ground commanders rely on UASs to provide timely, relevant, and useful intelligence without the lengthy processing and dissemination associated with the Air Force's centrally controlled, theater-wide assets. The Army's Training and Doctrine Command noted that the joint (CAOC [combined air operations center]) solution to meeting the high demand for these low-density assets has been ineffective, arguing against relying on the JFACC [Joint Force Air Component Commander] for UAS coverage. (Buchanan, 2010, p. 7).

JP 3-55.1 reflects a more service-centric notion of jointness.[3] When it was written in 1993, UASs were primarily thought of as reconnaissance, surveillance, and target acquisition (RSTA) assets. As envisioned then, mission planning was to be "based on the requirements of the supported unit," with due consideration for "airspace management conflicts" (JP 3-55.1, 1993, p. II-9).

Over time and with practical experience in Iraq and Afghanistan, the utility of UASs has expanded from consisting primarily of ISR to include armed surveillance and armed overwatch (e.g., for movement support and area security), targeting, communications support, attack, strike, and engagement; in the future, it will further expand to encompass logistical support and casualty evacuation missions. This expanded range of missions presents a challenge that was addressed in 2011 with the publication of a common set of multiservice TTP (MTTP).[4] Each of the four services has agreed to incorporate the MTTP into its respective training. The agreed-to MTTP document requires that

---

[3] JP 3-55.1, 1993, p. II-1, states that the

> [p]rimary mission of UAS units is to support their respective Service component commands as a tactical RSTA system providing the commander a capability to gather near-real-time data on opposing force position, composition, and state of readiness. However, as is the case with all assets and groupings within the joint force, the joint force commander (JFC) has full authority to assign missions to and task component UAS to conduct operations in support of the overall joint force.

[4] Reflecting the multiservice, rather than joint, nature of the document, each service has assigned its own number to it: Army Tactics, Techniques and Procedures 3-04.15, Marine Corps Reference Publication 3-42.1A, Navy Tactics, Techniques and Procedures 3-55.14, and Air Force Tactics, Techniques and Procedures 3-2.64. From here onward in this report, we will refer to the document simply as "the MTTP."

> group 3-5 UAS operations will be coordinated with the ACA (airspace control authority) and included in the ACO (airspace control order), SPINS (special instructions) and the ATO (air tasking order) in order to separate UASs from manned aircraft and to prevent engagement by friendly air defense systems. (MTTP, 2011, p. 12)

Table 3.1 describes the five UAS groups. The TTP for UASs that the services agreed to recognize that UASs may either be controlled by a centralized or joined C2 node (MTTP, 2011, p. 15) or be operated independently and that the service component commander may retain operational control of UASs (MTTP, 2011, p. 16), thereby avoiding the doctrinal issue that divided the Army and Air Force. In general, UAS control during the execution of an operation should be at the lowest tactical level to streamline the decision time lines and thus optimize responsiveness. The MTTP provides vignettes to help supporting and supported units understand how to incorporate and use available UAS assets more effectively. It should also be noted that no distinction is made between organic and nonorganic UAS support (MTTP, 2011, p. 40).

At least for the foreseeable future, each service will continue to field UASs and must bring to the fight forces fully capable of operating these aircraft and integrating their capabilities into joint operations. While some overarching doctrinal issues remain unresolved, the common procedures the services have agreed to and train for will remain the foundation on which UAS operations must be built, just as they are for military operations more generally. In this way, rather than being the foundation for UAS training, joint training is the apex of such training, which must be firmly grounded on the UAS training each service gives both its supported units and the supporting UASs.[5] This is the main focus of this report.

The UAS Training Concept: A General Notion for UAS Training

The contribution that UASs make to the modern battlefield results from a synergy among the platforms themselves, those who operate the platforms, and those who incorporate the UASs into the battle. Combat power cannot expand unless all three work together. A platform flown

Table 3.1
UAS Tiers by Service

Group	Capability	Examples
I	Hand-launched, self-contained, portable systems employed for a small unit or base security.	RQ-11A/B Raven
II	Small to medium in size and usually support brigade and intelligence, surveillance, reconnaissance, and target acquisition requirements.	ScanEagle
III	Operate at medium altitudes with medium to long range and endurance.	RQ-7 Shadow
IV	Relatively large UASs that operate at medium to high altitudes and have extended range and endurance.	MQ-1 Predator
V	Large, high-altitude, long-endurance UAV platforms	RQ-4 Global Hawk

SOURCE: DoD, 2011, pp. D-2 and D-3

---

[5] The terms *supported unit* and the *supporting UAS* are basic constructs of MTTP, 2011, p. 9.

by a skilled pilot does not add anything to the battle unless those on the ground who are engaged know how to exploit the information and other support the UAS provides. Similarly, if the pilot and payload operator cannot respond to the ground commander, the support they provide will not be what the commander needs. Only when all three work together will UASs be effective as a force multiplier. This can be achieved through training and practice that brings the platform, the operator, and the users together, starting at home station and progressing through ever-more-complicated exercises at the major training centers, such as the National Training Center (NTC) at Fort Irwin, California; the MCAGCC at Twentynine Palms, California; and the Joint Readiness Training Center at Fort Polk, Louisiana. Accordingly, how to achieve this synergy is the fundamental organizing concept for UAS training.

Let us be clear. The organizing concept developed here is not built around the training of a UAS pilot or payload operator per se. Their proficiency is certainly a necessary condition for UAS operations but is not, by itself, sufficient to enhance combat power. Enhanced combat power is achieved through the synergistic process that brings the platforms, the operators, and the users together. It starts at home station with small UASs and builds through the echelons of the ground forces as larger UASs are integrated into the combat operations of higher-echelon formations. Properly done, the maintenance or refinement of proficiency levels for UAS pilots and payload operators will be achieved incidental to the training that should take place with ground forces. But even before training can start, the services must decide how these systems will be used and how they will be incorporated into the force.

A Concept of Operations for UASs Is a Prerequisite for Training

As noted in the previous chapter, UASs are a classic disruptive technology. Integrating them into existing military operations is still a work in progress. This creates a problem for those charged with building training programs because, before training can start, the services must decide how these systems will be used and how they will be incorporated into the force. The services are still learning how best to use these new systems and the capabilities they provide. Each service is doing it differently.

Army

Initially, the UAS was thought of as just another platform for RSTA sensors, and the intelligence community was the proponent in the Army. In 2003, proponency for UASs transferred from Army Military Intelligence to the Army Aviation Branch. Since then, the UAS role in the Army has been changing from the relatively passive intelligence-gathering mission to the more active scout-reconnaissance and attack missions. One consequence of this transition is that UAS operators frequently team with manned aircrews to perform the scout-reconnaissance role. However, UAS operators and manned aircraft crews are currently separate communities and undergo separate programs of training.

At another level, the Army is experimenting with how to support the Raven, a small, hand-launched UAV. There is no dedicated military occupational specialty (MOS) for Raven operators. Being a Raven operator is an additional duty that requires additional training. Master Raven Training, a three-week train-the-trainer course, takes place both in the classroom and in the field. Students are taught how to maintain and operate the equipment and how to judge when and when not to fly. The plan calls for the trainer to return to home station and train squad-level personnel to operate the Raven.

Navy

The Navy has a different scheme for integrating the Fire Scout UAS and Broad Area Maritime Surveillance (BAMS) into the fleet. Typically, maintainers and pilots for Fire Scout have already gained technical training on manned helicopters. The Fire Scout–specific training is six weeks long and takes place at the Fire Scout Training Center at Naval Air Station Jacksonville, Florida (U.S. Army Maneuver Center of Excellence, undated).

The Navy's version of the Air Force's Global Hawk is the MQ-4C Triton, developed under the BAMS program. While the Navy continues to fly it to refine TTP for use in a maritime environment, the concept is for it to be integrated into active Navy maritime patrol units to complement the P-8 Poseidon, Boeing's 737–based multimission maritime aircraft, a traditional manned aircraft.

Air Force

The Air Force has yet another concept of operations that emphasizes remote split operations and reaches back to the Global Operations Center at Creech AFB, Nevada. Today, six operations centers in the continental United States (CONUS) support five launch-and-recover units in theater. Integration with tactical units is through tactical air control parties (TACPs) at the division, brigade, and battalion levels, and often at lower levels when supporting special operations forces (SOF). The TACPs coordinate directly with the Predator operations center via ultrahigh frequency radios on the aircraft or via satellite communications. As with other Air Force assets (and many air assets of other services), UAS assets are assigned to support ground commanders through the ATO. In the Air Force, the UAS pilots and payload operators are commissioned officers.

Marine Corps

The Marine Corps has dedicated UAV squadrons located at the MCAGCC at Twentynine Palms, California, and recently at Camp Pendleton, California. In the Marine Corps, enlisted Marines are responsible for almost every facet of the mission, from flying the aircraft and operating the payload camera to takeoffs and landings.

Summary

Even as the services are experimenting with the best way to use UASs and to integrate them into their forces, the overall purpose for UAS training should be clear. It is to increase combat power through the synergistic process that brings the platforms, the operators, *and the users* together. Putting these pieces together is the primary theme of the training framework we present.

Training Framework for Tactical UASs

The focus of the UAS training framework we offer is the enhancement of combat power, which starts with well-trained individuals who learn to work as a unit and ultimately leads to units that coordinate their efforts to produce military force. As we have seen, however, each service uses UASs differently.

Moreover, the great variety of UASs adds to this complexity. UASs are classified into tiers. Military planners designate the various individual aircraft elements in an overall usage plan using a tier system. The tiers do not refer to specific models of aircraft but rather to roles the aircraft typically fill. Table 3.1 shows the UAS groups and their roles and offers the examples of systems assigned to them.

The following discussion of training frameworks refers primarily to tactical UASs and not such Tier V systems as Global Hawk and its naval variant, the BAMS-D Triton.

The Training Framework for Ground Combat: Army, Marine Corps, and Air Force

The pyramid shown in Figure 3.1 represents our idea of the training framework to enhance combat power for the Army and Marine Corps, integrating organic UASs and Air Force UASs, when available, in support of ground forces. It is a graphic representation of the synergy achieved through training and practice that brings the platform, the operator, and the users together, starting at home station and progressing to higher echelons and ever-more-complicated battle exercises. The services have committed to align their training with the agreed-to MTTP. As with any good framework, the UAS framework must have a strong base. The base of this framework, represented by the lowest level of the pyramid, includes the initial and individual skills training of those who operate, maintain, and use the systems. The higher levels of the pyramid roughly align with the echelons of ground forces from the platoon or squad level through the company level and moving up to battalions, then to brigades. Each level incorporates the C2 of subordinate units to create combat power that is more than the sum of its parts. A battalion is more than just a collection of companies, each of which is more than just a collection of platoons and squads. In creating this combat power, the battalion utilizes UASs that are not normally available to its subordinate companies. In this way, the battalion and its UASs are "combat multipliers." This increase of combat power is repeated as we move up the pyramid to higher echelons each supported by the UASs that are organic to that

Figure 3.1
UAS Training Framework for Army, Marine Corps/Air Force Joint Combat Operations

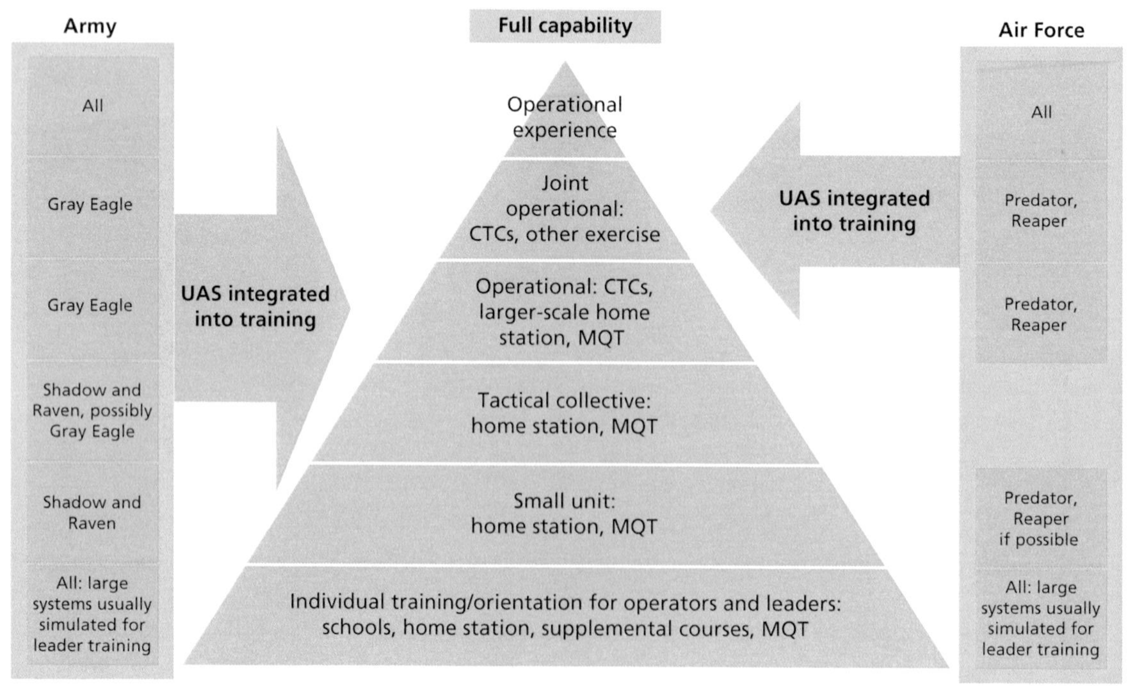

RAND *RR440-3.1*

tier. The two different-sized arrows, coming in from the left for the Army and from the right for the Air Force, show the specific UASs and the echelons they support.

In our concept of the training framework for UASs, the ground force systems—those of the Army and Marine Corps—and Air Force systems are "stacked" along the sides and paired with the level of the pyramid they support. The entries at each level of the pyramid identify the type of training that needs to be accomplished and where that training takes place. For example, small-unit training, at the lowest level of the squad and platoon, incorporates the Raven UAV and is accomplished at home station. Company training, which generally entails C2 of four platoons, is also accomplished at home station. The individual platoons may each be using their own Ravens, and the company commander may have access to the information from a Shadow UAS.

This process is repeated at home station with companies being brought together under the C2 of the battalion commander and his staff. During a battalion exercise, the battalion commander will generally be able to task the Shadow UASs either individually or as part of the combat support from assigned Army aviation. In some cases, this may include information provided by the Gray Eagle UAS. In some instances—these have historically been very rare—available Air Force UASs, such as Reaper or Predator, may support home station battalion exercises.

Brigade-level training can be carried out at some home stations, but the maneuver combat training centers (CTCs) generally provide a better venue. All available Army UASs will support these training exercises, from the small Ravens to the larger and higher-flying Gray Eagle, when available. In addition, it is at this level and during these exercises that joint Army–Air Force support is usually integrated to include close air support (CAS) and support from Air Force UASs.

Not shown on the pyramid is the training needed to meet the certification requirements of UAS operators. Such requirements will normally be met incidental to their training with ground forces. For example, Shadow operators need to operate their equipment 4.5 hours per month to maintain their certification. It is expected that they will be spending more than that supporting ground units during home station training. Operators not actively engaged with the training of ground units can achieve the needed proficiency training using the same training equipment as during their initial skill training.

The Training Framework for Fleet Operations

A similar pyramid also represents our framework for Navy UAS training (Figure 3.2). It differs from the previous pyramid in that the Navy does not have a hierarchy of systems introduced at higher echelons of command. At the tactical level, the only UAS the Navy has is Fire Scout, which will be introduced into fleet operations following a similar path as discussed above. As with land forces and the systems that support them, combat proficiency is generated through the synergy of the platform, operators, and staffs that incorporate the UASs into the battle planning. Initial operator training is the base of the pyramid and takes place in the schoolhouse in the vicinity of the homeport. Operational training occurs under way in fleet operating areas. The Navy trains continuously, and the training continues through the operational use of UASs when the units are deployed.

UAS operators and maintainers require initial training to obtain the skills needed for operational effectiveness. As individuals become qualified as UAS operators and maintainers, they undergo team training to achieve mission area certification. For example, UAS operators

Figure 3.2
UAS Training Framework for Navy Tactical UAS

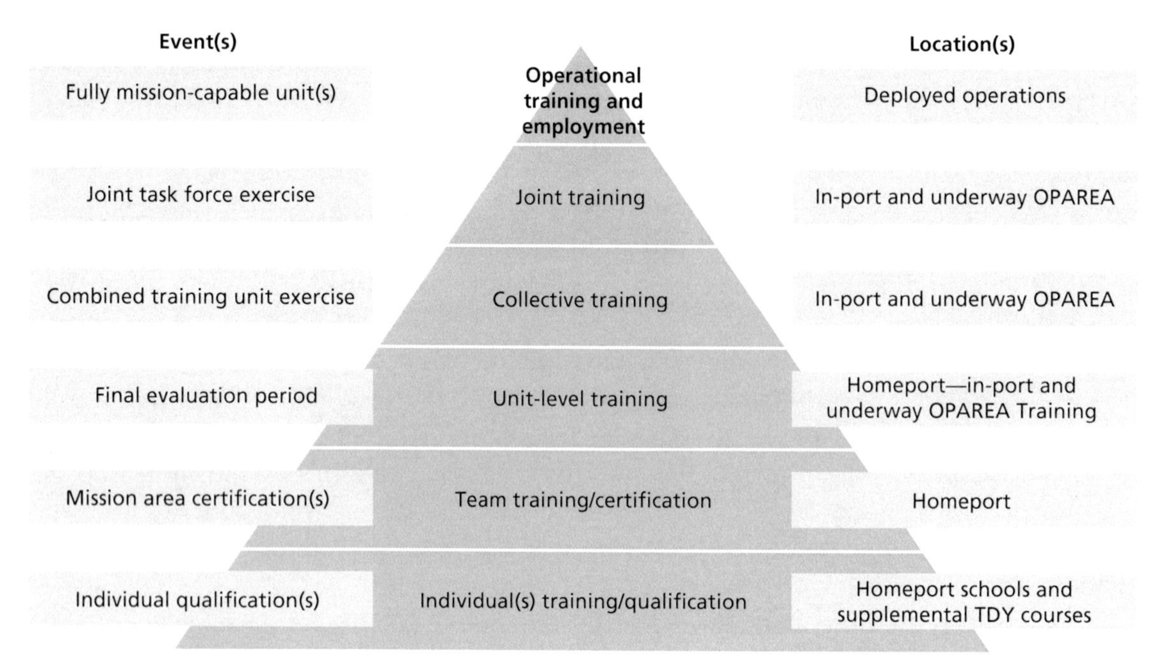

RAND *RR440-3.2*

may be linked to and integrated with an antisurface warfare mission team to support certification for the surface warfare mission area. For ships, unit-level training occurs both in port and under way and culminates in a final evaluation period. During that evaluation, the immediate superior in command evaluates the ship in all mission areas to determine that it is ready for advanced phase training. The final evaluation period comprises in-port and underway training events geared to assess the unit's ability to operate independently.

Collective training occurs after the final evaluation period, when the ship begins to operate with other units and staffs. During this time, the ship conducts operations with and becomes an integral part of a carrier strike group (CSG).[6] Collective training is a combination of in-port and underway training events and culminates in a combined training unit exercise evaluated at the fleet commander level. The Navy is in the process of developing its UAS tactics and procedures and plans to integrate UAS training into predeployment preparations.

Deployed naval ships and units must also operate in a joint environment. After completing a combined training unit exercise, a deploying CSG conducts a final exercise with elements of other services (e.g., Air Force aircraft) to demonstrate its ability to communicate and operate effectively in joint warfare. Joint training is conducted both in port and under way. An underway joint task force exercise is the culminating event, with Navy units operating with forces of the other services; on satisfactory completion, units are deemed ready to deploy. Units then continue the operational training and employment of their systems en route to their deployed operational areas.

---

[6] While we address surface units as being part of a CSG, surface units can also deploy independently, with other ships, or with an expeditionary strike group.

Navy units deployed to 5th Fleet do have experience with BAMS–Demonstrator (BAMS-D), the predecessor to the Triton, and a variant of the Air Force's Global Hawk. Despite being a demonstrator, BAMS-D was deployed to meet surveillance needs in the 5th Fleet. Navy strike groups operating in 5th Fleet were provided with imagery and data feeds from BAMS-D and used these data to build a common operational picture, which increased situational awareness. The BAMS-D was integrated into actual fleet operations and provided near-real-time updates for surface surveillance. Units learned to use BAMS-D "on the fly," by processing, exploiting, and disseminating imagery and sensor data; building the surface picture; and increasing the positive identification of surface contacts in the area of operations.

As UAS assets are fielded, they will be incorporated to a greater degree into predeployment training and increase opportunities to train and hone the tasking, collection, processing, exploitation, and dissemination of UAS sensor data.

Applying the Training Framework

The training framework presented here focuses on the synergy among three key elements: the platform itself, those who operate the platform, and those who incorporate UASs into the battle. The framework emphasizes using UASs to enhance warfighting proficiency. It provides a useful tool for organizing and evaluating training options and assessing the current state of training and plans for improvement. As discussed in the next chapter, we used the framework to assess the state of UAS training, as it existed in 2012.

CHAPTER FOUR

Assessment of the State of UAS Training in 2012: Service and Interoperability Training and the Role for Simulators

For more than a decade, UAS training has concentrated on preparing UAS pilots and payload operators to support ongoing operations in Afghanistan and Iraq. This remained true in 2012 and 2013, and the services continue to evolve how best to use these systems and integrate them into ongoing operations. With the operations in Afghanistan winding down, the services must now integrate all facets of UAS training, including the infrastructure for training addressed in Appendix C, into a largely CONUS-based inventory of UASs that continues to grow, even as real-time operations continue for some systems.[1] Moreover, operational demands and technical innovations will continue to evolve and affect training and organizational requirements. As a result, UAS training and training support programs must be viewed as works in progress.

In assessing the current state of UAS training, using the UAS training framework discussed in the previous chapter, we view proficiency in flying UASs not as an end in itself but as only a means to an operational end: accomplishment of a military mission. As noted, this is achieved through the synergistic use of UASs with the forces they support, in ways similar to those in which the timely integration of intelligence or the delivery of fire support is accomplished in combat operations. This convergence requires full integration of UASs into the collective training of operational forces, attendant training and education for leaders and staffs, and the integration of lessons learned from previous and ongoing operations into education and training programs.

The preceding should by no means be construed to understate the need for training those directly involved in the operation of UAS—among them pilots, crews, support personnel, and those who maintain the aircraft and communications systems. Such training, both to develop and to maintain proficiency in the operation of UAS, is critical for the overall success of a UAS training strategy. But proficiency in these areas is not in itself sufficient. Full understanding of UAS integration, from the smallest to the largest systems at every echelon, as outlined in the previous chapter, is also necessary.

To be clear, members of the UAS community not only must understand how to operate their systems—a challenge largely being met successfully—but must also be familiar with the fundamentals of the operations they support and how their capabilities contribute to the successful accomplishment of the mission. Similarly, battle staffs must fully understand the many capabilities of UASs, how these capabilities supplement or complement other capabilities available to the joint force, and how best to integrate the capabilities into their operations.

---

[1] The Air Force UAS community believes high demand for its systems—Predator and Reaper—will persist, with requirements to support other kinds of demands increasing as demands for support of ground maneuver operations decrease.

This chapter assesses the state of training for UASs across the ground, air, and sea forces. It discusses information, observations, and insights gained from a wide variety of sources but mainly through visits to a select number of service bases in 2012. During these visits, the RAND team talked with operators and support personnel; those who train operators and associated members of the team; those who train and observe the forces who employ UASs; staff elements concerned with the planning and resourcing of training, both for UAS operators and for the force more generally; and those concerned with the ongoing development of doctrine and concepts of operations for UASs, including their integration into operations. What follows is a compilation of the observations and insights from these individuals and our interpretation of the implications for the specification and resourcing of training strategies.

Ground Forces

Ground force operations of the Army and Marine Corps, even at low echelons, can be highly complex, and the complexity increases significantly at higher echelons. Commanders at higher echelons not only have to coordinate and use the information and capabilities subordinate units provide but also have to manage and integrate information and capabilities from a widening variety of other sources available to them. Adding the integration of UAS capabilities into the operational command further increases its complexity and thus the complexity of associated training management processes. Evidence from both training exercises and actual operations in combat theaters shows that, when properly managed, UASs provide significant new capabilities that commanders at every echelon can effectively employ. UASs truly have the potential to be force multipliers—but more so when employed by users fully competent in integrating their capabilities with those of the rest of the force and its supporting elements.

As shown in the training framework presented in Chapter Three, the proper integration of UASs to create combat capability starts with the training of UAS operators and support personnel and progresses through the integration of UAS operations with supported forces.

Training for Operators and Support Personnel

In general, the RAND team found that qualification training in the Army and Marine Corps for designated UAS-specific MOS is well established. In the Army, positions are filled by enlisted soldiers. The ranks of operators typically range from private first class to master sergeant (E-3 to E-8). The Marine Corps is organized much like the Air Force, with a UA commander who is a company grade, qualified aviator or aviation C2 officer on his or her second or third tour (O-3/O-4) and a mission payload operator (MPO) who is typically a corporal through master sergeant (E-3 to E-8).

Army Air Vehicle and Sensor Operators for Hunter, Gray Eagle, and Shadow UASs

Training for Army UAS operators—pilots and payload operators—is carried out in two phases: The first, at Fort Huachuca, consists of a common core course for all operators of UASs not launched by hand.[2] This phase lasts nine weeks and two days and has five sections. (See Figures 4.1 and 4.2.)

[2] The 2nd Battalion, 13th Aviation Regiment, which was previously called the Unmanned Aircraft Systems Training Battalion, conducts the training. See Appendix D for RAND's "military value analysis of training bases" assessment of training at Fort Huachuca, Arizona.

Figure 4.1
Army UAS Air Vehicle Operator and APO Training Pipeline

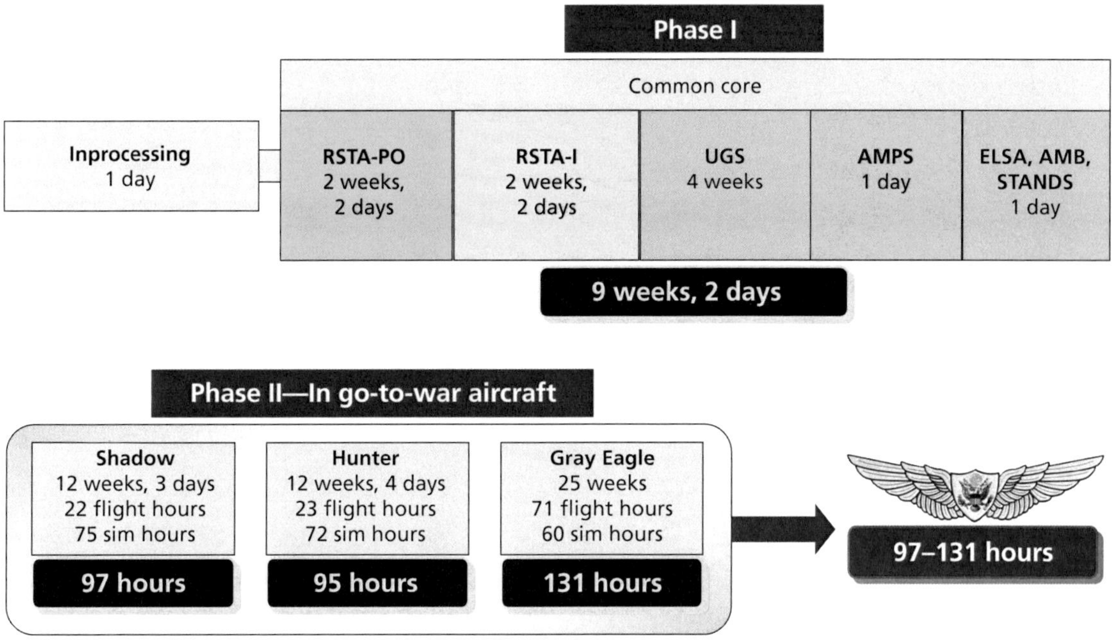

Phase I

Common core

Inprocessing
1 day

RSTA-PO
2 weeks,
2 days

RSTA-I
2 weeks,
2 days

UGS
4 weeks

AMPS
1 day

ELSA, AMB,
STANDS
1 day

9 weeks, 2 days

Phase II—In go-to-war aircraft

Shadow
12 weeks, 3 days
22 flight hours
75 sim hours
97 hours

Hunter
12 weeks, 4 days
23 flight hours
72 sim hours
95 hours

Gray Eagle
25 weeks
71 flight hours
60 sim hours
131 hours

97–131 hours

RAND *RR440-4.1*

After completing Phase I, the operators continue to Phase II. This section is unique for each airframe and consists of both simulation and flight hours. It typically takes 12.5 weeks for Shadow and Hunter classes to finish and 25 weeks for Gray Eagle classes. After completing Phase II, the UAS operators earn their wings.

Maintenance for Hunter, Gray Eagle, and Shadow UASs

The Army UAS maintainers take the MOS 15E Repairer course from the 2nd Battalion, 13th Aviation Regiment (see Figure 4.3). As shown in Figure 4.4, the course lasts 17 weeks, and a new class of up to 16 students starts every two weeks. After completion, 15E personnel are qualified to maintain the Shadow. Maintainers must take additional training to earn additional skill identifiers and thus qualify to maintain the Hunter and Gray Eagle. The additional training takes ten weeks for Hunter and 18 weeks for Gray Eagle.

Army Raven Operators

Without a dedicated MOS, the development and maintenance of Raven operators is a unit personnel and training management issue. Assignment as a Raven operator is an additional duty. In the past, this has meant that many units did not have the required Raven-qualified personnel to employ the systems properly. The Army is well aware of this problem and, short of creating a dedicated MOS, has recently initiated a program to deal with the Raven training problem. The Army Maneuver Center of Excellence at Fort Benning manages a certification course designed to qualify selected soldiers as small UAS (SUAS) master trainers capable of conducting initial qualification training (IQT) and certification of new operators at home stations throughout the Army. This train-the-trainer course focuses on how to teach and manage an aircrew training program at home base and how to assist small-unit commanders in evaluat-

Figure 4.2
Shadow (top) and Gray Eagle UAVs

SOURCE: DoD.

SOURCE: U.S. Army via social media outreach.
RAND *RR440-4.2*

ing academic and flight instruction and managing SUAS accident prevention.[3] It is envisioned that new master trainers returning to their home stations will be able to develop and maintain

[3] The 15-day course includes classroom and hands-on training, with instruction on reporting procedures, fundamentals of instruction, semiannual evaluations, and familiarization with Raven A and B and digital downlink systems. At the conclusion of the training program, graduates should have the ability to evaluate and certify operators. See U.S. Army Maneuver Center of Excellence, undated.

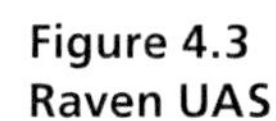

Figure 4.3
Raven UAS

SOURCE: U.S. Army via social media outreach.
RAND *RR440-4.3*

a sufficient number of operators within their units and manage a tracking program to ensure that their units have the required number of certified Raven operators.

Marine Corps Unmanned Aircraft Commanders—Pilots

Training begins with the Unmanned Aircraft Commander course, which last two weeks and is taught by the U.S. Army at Fort Huachuca. Career-level training takes place at the squadron level. Marine Aviation Weapons and Tactics Squadron 1, at Marine Corps Air Station Yuma, Arizona, provides advanced training (the Weapons and Tactics Instructor course).

Air Vehicle Operators and Mission Payload Operators

Training for air vehicle operators (AVOs) and MPOs is accomplished jointly with the U.S. Army at Fort Huachuca. USMC operators take the nine-week common course and the 12-week Shadow course. Career-level training takes place at the squadron level. Marine Aviation Weapons and Tactics Squadron 1 provides advanced training.

Figure 4.4
Army UAS Maintainer Pipeline

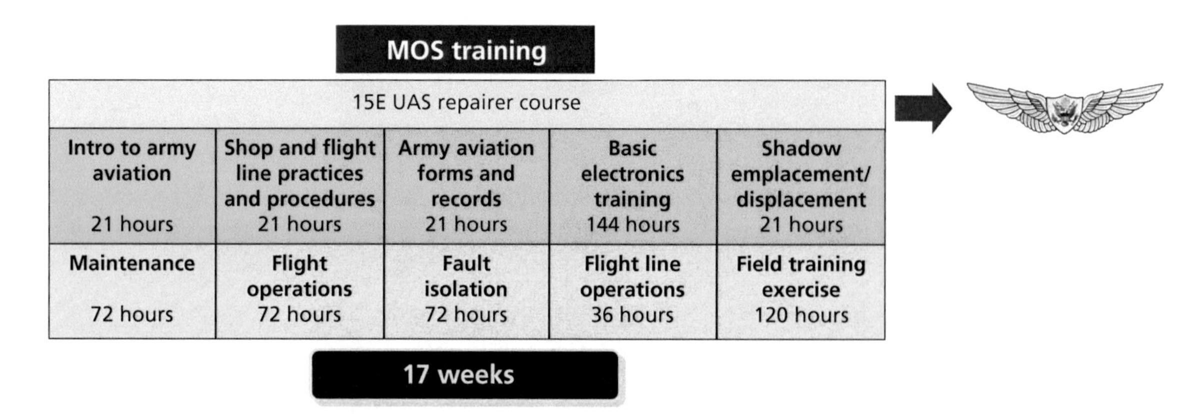

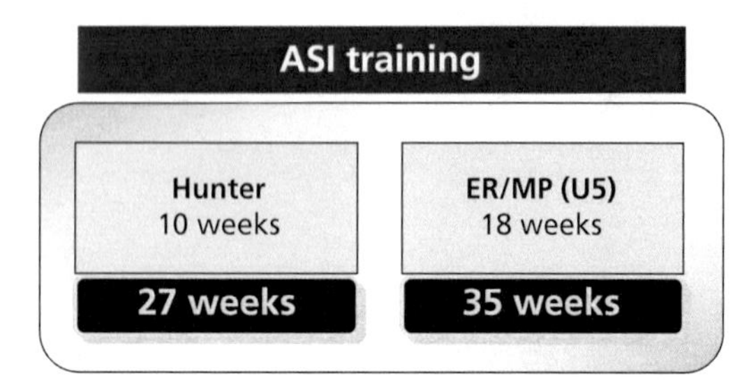

RAND *RR440-4.4*

Operational Training[4]

Larger UASs have dedicated MOS training, so operators and maintenance personnel arrive at home station having completed their IQT. To maintain their proficiency and certification, crews must log a minimum number of flying hours. The Army's Gray Eagle UAS has its own simulator capability built into the operating hardware. Currently, the Army's requirement is for 4.5 live flying hours per crew per month for the Shadow UAS. The Army has had difficulty maintaining sufficient numbers of Raven-qualified SUAS operators, in part because units do not always do well in tracking their qualified operators, maintaining their proficiency through hands-on work, and providing for their timely replacement when they move on. Discussions with Army trainers indicate that, despite the demands of deploying units and sustaining combat operations in Afghanistan, Army units are doing better at maintaining Shadow qualification. Nevertheless, not all units have been able to maintain high Shadow qualification rates: Some units arrive at the NTC with one-half or fewer of their operators qualified.

For these systems, the critical issue for the future will be securing enough flying time for UAS crews to train with ground forces. Marine Corps experience offers a good illustration of this issue. The Marine Corps has based Shadow squadrons at the MCAGCC, where there is adequate restricted airspace both to maintain crew proficiency and to fly in support of ground forces. However, both the ground trainers and squadron personnel the RAND team interviewed reported having to start what was essentially remedial training because the ground

[4] Joint Publication 1-02 defines *operational training* as "training that develops, maintains, or improves the operational readiness of individuals or units" (JP 1-02, 2013).

forces were not adequately trained at home station to integrate UAS capabilities into their operations. These students generally could not progress to the level of proficiency desired.

The foregoing suggests the biggest challenge in developing and implementing a training strategy for UASs is gaining and maintaining proficiency on the part of end users—the leaders and staff members who will integrate UAS capabilities into their operations.[5] Ground units that undertake joint training at CTCs are often unprepared to use organic and nonorganic UASs effectively, and thus tend to develop battle plans that underutilize them.[6] Further, their intelligence and operations staffs sometimes find themselves at cross purposes over employment of UASs, an issue that could be better dealt with during home-station training.

Despite all this, discussions with operators, users, trainers, and training managers suggest that properly employed UASs consistently prove to be valuable combat multipliers. One example was an Army unit training at the NTC, which had successfully employed its UASs as part of a manned-unmanned team. In this instance, the UAS had acted as a scout and observer as manned attack helicopters delivered fires on selected targets, all in support of a ground maneuver unit. Another battalion well versed in the myriad tasks associated with employing Shadows—communication of commander's intent and concept, priority and time sensitivity of information requirements, targeting priorities, use of the communications relay package, to list just a few—was far more successful in employing the Shadows supporting it than another unit was. In the latter unit, the UAS platoon was simply given "areas to look at," without specific guidance on what to look for, e.g., named areas of interest and targeted areas of interest. Interviewees told the RAND team that many UAS operators do not get enough opportunity to interact with either crews of manned aircraft or the elements of ground units—commanders and operations and intelligence sections—that they will support.

The latter problem, ground units' lack of experience in employing UASs, is all too common. It is also true for the USMC, whose units face considerable limitations on using UASs at their home stations or local training events.[7] Marine ground force units are thus typically not well versed in employment of UASs when they arrive at Twentynine Palms. Although they learn while there, many are still not fully capable when they deploy.[8]

Improvements in Facilities and Basing for UASs Will Improve Training Integration

To date, the Army has had reasonable opportunity to train at home station, and current programs to increase UAS facilities will appreciably add to this. Table 4.1 summarizes responses the U.S. Army Training and Doctrine Command (TRADOC) received from active, National Guard, and reserve bases in November 2012 concerning the availability of restricted airspace and resulting limitations on UAS operations stemming from lack of availability. The responses

[5] We refer to anyone who could benefit from the information a UAS provides as an *end user*. The MTTP calls end users *supported units*.

[6] *Organic* assets are ones owned by a unit, and *nonorganic* assets are ones that the commander of a unit does not own and thus does not have full control over.

[7] Recall that the Marines' Shadow squadrons are based at Twentynine Palms when not deployed.

[8] This brings up the question of whether we should expect units to continue learning sound tactics and procedures while they are deployed. They should, of course. The issue is how much; officials we have consulted have given us the sense that many units are still on the steep part of the learning curve when they leave the MCGACC. These units would be more operationally effective sooner in their deployments if they could be better trained to integrate UAS capabilities before they deploy.

indicate that, in many cases, relatively minor military construction projects are all that is necessary to expand training opportunities at home stations. In most locations, additional bed-down facilities will allow better use of existing restricted airspace to facilitate live training. In a few cases, however, the proximity of training areas to the Federal Aviation Administration (FAA)–controlled National Airspace System (NAS) remains a problem. The summary in Table 4.1 provides a snapshot of the kinds of issues that have already been raised and will have to be revisited at some point because of the ongoing pressure to reduce defense budgets.

Table 4.1
Select Summary of Army UAS Installation Survey as of November 21, 2012

Base/Commands	Answer To Question E: Does Available Restricted Airspace Affect/Limit UAS Operations at Your Installation?	Answer To Question J: What Solutions Would You Recommend for Unsolved Issues?	Shadow Strips Planned for FY 2013
Alaska (Forts Wainwright and Richardson)	No.	We have been able to accommodate all down range requests. However, the hangars are nonstandard and need to be evaluated for suitability as the Shadow and Gray Eagle platforms evolve.	
Fort A. P. Hill	During landings the tactical approach landing system orbit is outside the restricted airspace.		Yes
Fort Bliss	Our special use airspace is bisected by a major highway which requires a COA.	Yes, a Shadow Strip in the SUA.	Yes
Fort Bragg	Yes.	Yes. Primarily commercial power and hardened maintenance facilities. Temporary solutions are power generation equipment and tentage.	Yes
Fort Carson	Yes, limited by the volume of manned/unmanned platforms operating within the airspace.	Yes, requests for designated UAS facilities and airstrip for launches and recovery. Funding established to build a multiple platoon facility for UAS units.	Yes
Fort Drum	No.	No. We have no unresolved UAS issues.	Yes
Fort Hood	Not currently.		
Fort Knox	No.	Yes; the construction of a tactical UAS field support structure (build it).	
Hawaii	Yes, we have requested to use the Kahuku and East Ranges for flights but due to FAA restrictions, we cannot fly UAS/SUAS platforms in these airspaces.	No. We need more airspace to train in. Our current location is becoming less and less amicable to flight conditions due to public demands on land. In 2010, plans were drawn up to provide for a permanent UAS facility at Wheeler Airfield, but they were cancelled due to program funding.	

Table 4.1—Continued

Base/Commands	Answer To Question E: Does Available Restricted Airspace Affect/Limit UAS Operations at Your Installation?	Answer To Question J: What Solutions Would You Recommend for Unsolved Issues?	Shadow Strips Planned for FY 2013
Fort Lewis (Joint Base Lewis-McChord [JBLM])	1. Yakima Training Center (YTC): No not an issue. 2. JBLM: Yes, both the limited restricted airspace and the large number of installation units requiring restricted airspace, challenge UAS/SUAS operations.	1. YTC: Shadow Training Facility 2. JBLM: Construction of a second Shadow landing strip. 3. JBLM: Construction of any type of permanent, securable facility alongside the existing Shadow landing strip (location is outside JBLM's guarded area). 4. JBLM: Shadow Training Facility or TEMF.	
Fort Riley	Not currently. (See response to Q h & i.)	Nothing to comment on.	Yes
Fort Stewart	Not significantly. However, reductions in air traffic control manning are a concern. There are currently no significant restricted airspace challenges negatively affecting UAS operations.	Funding, we have a plan and the requisite site approvals, we just need funding support. We need the funding ($701,000) now for our third Shadow platoon hangar; we have the design ready to go and at contracting.	
USAREUR	Yes. 173rd Shadow out of Italy is currently based in Bamberg due to no restricted airspace to fly in Italy. Currently fly in restricted airspace in Grafenwoehr/Vilseck Training Areas.		
Camp Atterbury	No, both UAS facilities are within restricted airspace.		
Fort Chaffee	No restricted airspace limitations to TUAS flight operations.		Yes
Orchard Range	Yes, UAS operations are limited to using R-3203.	Fixed support facilities UAS operations. Facilities will be in Range Complex Management Plan.	
Fort Pickett	No, we have one of the largest SUAs in the Midatlantic.		
Camp Ripley	No, the restricted airspace at Camp Ripley (R-4301) provides enough area to conduct training for the current systems.	Yes. The current tactical strip will be removed for the construction of a MPTR. UAS commanders have requested a tactical strip for replacement that includes hard-stand buildings and electrical power. The Combat Readiness Training Center currently has a location selected that has power in close proximity to the existing tactical strip. Camp Ripley is one of the sites to receive $730,000 for construction of a UAS strip in the down-range area in FY 2013.	Yes
Camp Roberts	No.	Funding.	Yes

Table 4.1—Continued

Base/Commands	Answer To Question E: Does Available Restricted Airspace Affect/Limit UAS Operations at Your Installation?	Answer To Question J: What Solutions Would You Recommend for Unsolved Issues?	Shadow Strips Planned for FY 2013
Fort Dix	Yes.	We have accommodated all downrange requests. However, the size restriction of the UAS in R-5001 limits operation of large UAS operations.	
Fort Hunter-Liggett	No, we use R-2513 airspace.	None.	
Fort McCoy	Yes, Current coutilization of Young Air Assault Ship, a 6,250-ft. C-17/C-130 landing strip in R-6901 B. The UAS operations at this landing strip affect Sparta/ Fort McCoy airport operations.	Yes. Hangers for Shadows and permanent facilities with commercial power, and asphalt. Standard design and funding required.	
Fort Campbell	Fort Campbell currently has two approved sites for Shadow Flight operation; both are located among numerous field artillery firing points, which requires de-confliction during launches and landings because the firing points have to check fire. As a result, UAS flight training is significantly reduced. However, an additional UAS strip is being constructed to alleviate this problem.	Yes. Permanent facilities with commercial power and utilities.	

SOURCE: Installation UAS survey responses from TRADOC Capability Manager—Live, as of November 21, 2012.

Key Areas of Concern

As the war in Afghanistan draws to a close, the services must pay close attention not only to the training of UAS crews and support personnel but also to the routine exercise of these systems as part of combined arms training at home station and at the CTCs. This will require attention to a number of issues.

Beddown and Related Support Facilities

To facilitate the efforts described above, DoD should support current and future programs to develop ranges and beddown and support facilities similar to those in the Army's current programs. The Army has been implementing a robust program to establish Shadow beddown facilities—runways, shelters, and attendant support structures—on many of its installations, both active and National Guard. This will enable more opportunities both for UAS operators and for end users.

Airspace Considerations

As the responses reported in Table 4.1 suggest, the unique limitations of UASs pertaining to operations in the NAS can make integration into home station training difficult. Currently, UASs can operate only in restricted military airspace or, in the NAS, with certificates of authorization (COAs) from the FAA. In some cases, it should be possible to increase UAS training with ground forces (and with collocated manned aircraft units) without impinging in any significant way on the NAS. Although this will certainly require better coordination of restricted airspace at individual installations, better availability of restricted airspace will make better access to the NAS less of a key element in supporting training strategies, at least as they pertain to home-station or local training of ground forces. However, less-cumbersome procedures for NAS access will, among other things, facilitate opportunities for units to employ larger UASs, such as Gray Eagle, Predator, and Reaper, in home-station training events.[9]

Air Force

The Army and the Air Force have significantly different concepts for employment of UASs. The Air Force uses a remote split operations concept to employ its UASs. Under this approach, the Air Force flies UASs from home station using satellite or other relay links. It maintains only small deployed footprints at consolidated operating locations to launch, recover, and maintain the aircraft.[10] This approach enables the deployment of the vast majority of its systems, yet the majority of the personnel who operate Air Force UASs do so from home station and require no reconstitution following deployments. By using remote split operations, 85 percent of the Air Force's UASs would be forward deployed for combat operations. The remaining 15 percent would support training at each operational base. The Air Force employment concept, however, does not facilitate integrated operations with Army tactical-level units during training at Army home stations. Once deployed, Army brigades, battalions, and SOF must learn to work with Air Force TACPs to coordinate UAS operations.

Training for Pilots, Operators, and Support Personnel

The Air Force uses a crew of one pilot and one sensor operator to man each UAS mission.[11] The pilots are officers and are either rated pilots who graduated from undergraduate pilot training or UAS-only pilots who graduated from a new program called undergraduate remotely piloted aircraft training.[12] UAS-only pilots have the 18X Air Force Specialty Code (AFSC) and will be discussed in more detail later. Following the structure of the training pipelines for other platforms, the training is broken into three phases: IQT, mission qualification training (MQT), and crew mission ready training. Air Education and Training Command conducts IQT and MQT, with the latter taking place in a formal training unit (FTU).

---

[9] In some cases, basing for the larger systems, especially Gray Eagle, will be contiguous with the training areas in which they are needed; Fort Hood, Texas, is an example. But this will not always be possible. The larger systems require longer and better runways and larger and more-sophisticated beddown and support facilities. It may be more cost-effective to move these UASs to support training where they are needed, rather than establishing beddown facilities in a large number of additional locations.

[10] The 432nd Wing at Creech AFB operates MQ-1 Predators and MQ-9 Reapers from seven forward bases overseas. Maintainers at forward bases serve four-month rotations in theater.

[11] Many more people are involved in RPA operations, including the mission intelligence commander. Also, there is a weapon school course for advanced training of RPA operations. For this research, we were primarily interested in the initial training of pilots and sensor operators.

[12] Some Air Force navigators who had commercial instrument qualifications have been allowed to train as pilots for the RQ-4. This avenue for sourcing RQ-4 pilots closed in spring 2011.

Pilots

Reaper and Predator

In response to the need to reach and sustain the manpower to fulfill Reaper and Predator requirements, the Air Force initiated a new program to produce RPA pilots.[13] Currently, Reaper training takes place at Hancock Field Air National Guard Base in Syracuse, and Predator training takes place at Holloman AFB. New RPA pilots have attained the foundational skills to conduct RPA missions and are qualified to operate in the NAS or International Civilian Aviation Organization airspace. Figure 4.5 outlines the training pipeline. Figure 4.6 shows the UASs.

Global Hawk

Currently, most Global Hawk pilots are traditional Air Force pilots who volunteered for a tour in the Global Hawk (see Figure 4.7), a majority of whom are qualified to fly multiengine transport aircraft. Prior to 2011, navigators who also had commercial pilot licenses and had completed commercial instrument qualifications were allowed to become Global Hawk pilots; however, the Air Force has not decided whether to retain navigators who are currently Global Hawk pilots in the community or return them to navigator duties. The third source of Global Hawk pilots is the 18X career field, UAS-only pilots who have graduated from undergraduate remotely piloted aircraft training. After finishing the three preparatory courses, candidate Global Hawk pilots head to the RQ-4 FTU, the 1st Reconnaissance Squadron at Beale AFB.

Figure 4.5
Air Force UAS Pilot and Sensor Operator Pipeline

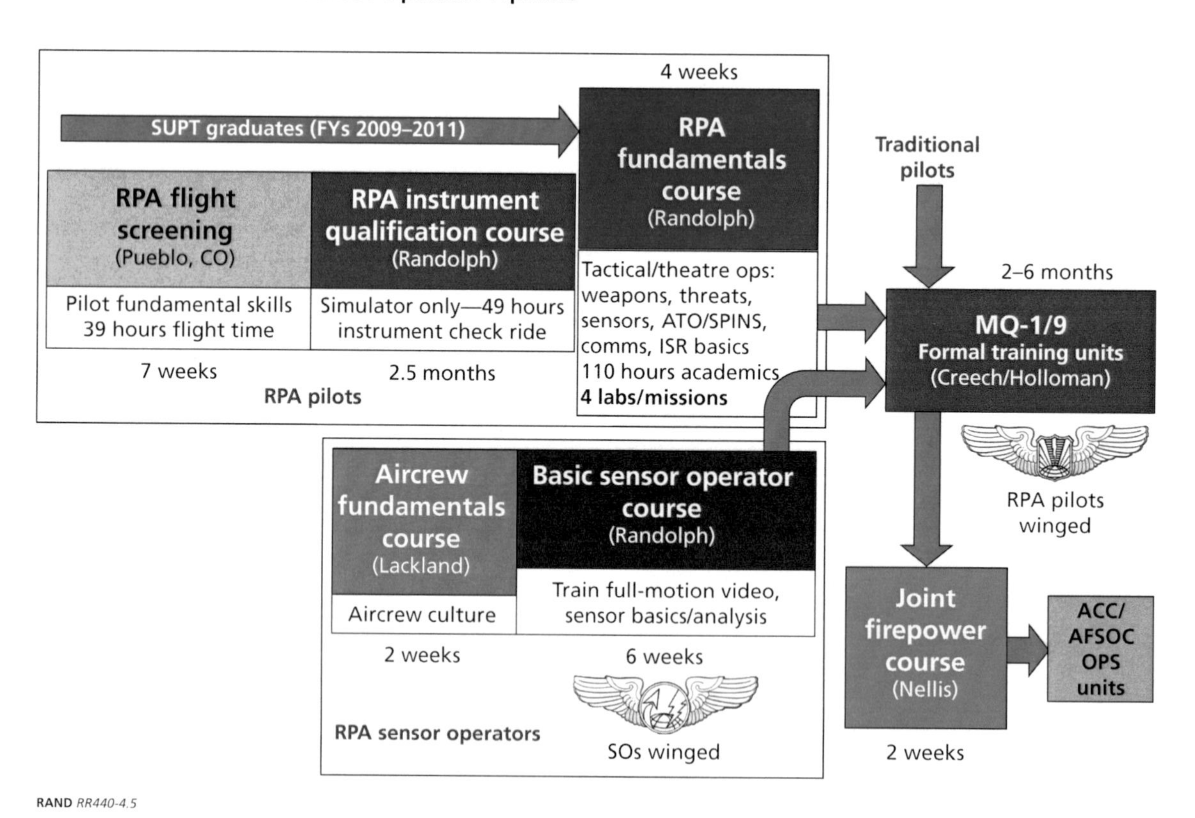

RAND *RR440-4.5*

[13] The Air Force uses the term *RPA* to refer to unmanned aircraft systems. The accepted joint terminology is *UAS*.

Figure 4.6
Reaper (top) and Predator UASs

SOURCE: U.S. Air Force.

SOURCE: DoD.
RAND *RR440-4.6*

Once at the FTU, pilots are trained first on simulators designed for pilots with an instructor pilot. Then, they participate in on-the-job training, flying live missions with an instructor pilot in the mission control element.

Figure 4.7
Global Hawk UAS

SOURCE: DoD.
RAND *RR440-4.7*

Sensor Operators

Reaper and Predator

Enlisted sensor operators for Reaper and Predator are either trained into a new career field or cross-trained from another AFSC (1U00X). After graduating from the Air Force's basic training or being reclassified, they attend a two-week course at Lackland AFB, Texas, with other aircrew AFSCs. Then they go to Randolph AFB, Texas, for a six-week basic sensor operator course. After completion, they go to the appropriate FTU for the system they will be flying.

Global Hawk

Sensor operators for the Global Hawk are trained differently from those for other UASs. Sensor operators for the Global Hawk come from the intelligence imagery analyst career field (1N1XXs). They receive MQT at Beale AFB, where they learn to use the Global Hawk system. After one tour as sensor operators (approximately three years), they move on to other imagery analyst assignments; therefore, it is difficult to maintain expert sensor operators for the Global Hawk.

Maintenance

Training for those who maintain Air Force UASs takes place at three locations.[14] The fundamentals are taught at the maintenance schoolhouse at Sheppard AFB, Texas. Initial skills training and advanced skills training take place at the unit.

---

[14] Contractors provide certain UAS maintenance, for example, depot-level maintenance.

Operational Training

Given the emphasis on current operations and the Air Force's remote split operations concept for employing UASs, few opportunities exist to support ground troops either during home-station or CTC training. In general, with a few limited exceptions, Air Force UAS crews do not train in exercises or other training events in CONUS and do not support training in CONUS of Army or Marine Corps units at either their home stations or at the CTCs. Air Force UAS crews do conduct operational missions in theater, and these missions frequently are the chief or even the only way deployed Army and Marine Corps units get an opportunity to work directly with Air Force UAS crews. Thus, most elements of the UAS force in theater often have no experience working together before they have to do so in combat operations.

The result of not working with ground troops either at home station or the CTC is that ground units generally do not integrate Air Force UAS capabilities into their planning. RAND's conversations with the training community, including trainers at the NTC, as well as what officials told the GAO, confirmed that the

> effective integration of UAS in training exercises, like the integration of other types of joint air assets, depends on the priority that ground units place on developing training objectives that require the participation of joint air assets and their ability to plan for the use of these assets in the exercise. ... [As a result], Army combat brigades often focus UAS training objectives during exercises on integrating their Shadow UAS and do not emphasize planning for and employing Air Force UAS.[15]

---

[15] Independent RAND team visits also confirmed what the Air Force told the GAO, that

> unmanned aircraft are deployed to support overseas operations except for those that are supporting the initial training of UAS personnel or the testing of aircraft." These officials [from the 432 Wing] stated that in the event that additional aircraft were made available, the wing's personnel levels are insufficient to support additional training events because the unit does not have adequate personnel to support projected operational commitments and greater numbers of training exercises. Second, Army and Air Force officials told us that when Air Force UASs are at the training center, these aircraft are not always available to support ground unit training because a considerable portion of the UAS flight time is dedicated to accomplishing Air Force crewmember training tasks. Officials told us that the Army and Air Force have reached an informal agreement to allot about half of the time that an Air Force UAS is flying at the training center to support Army ground unit training objectives and the other half to accomplish Air Force training tasks. Air Force officials pointed out that although they try to align their crewmember training syllabi with ground unit training objectives at the National Training Center, training new personnel to operate these aircraft is their priority. Third, UASs may not be available during certain hours to support ground unit training, which during exercises goes on 24 hours a day. For example, Predator UASs from the California Air National Guard are available to support ground units only during daylight hours. To travel to the training center, these aircraft must pass through segments of national airspace that are not restricted for DOD's use and therefore must rely on a ground-based observer or on chase aircraft to follow them to and from the training center. Because of this reliance on ground or airborne observers, flights to and from the training center must be accomplished during daylight hours and are necessarily more expensive as well.
>
> As a result of the limited number of unmanned Air Force assets that are available to support ground unit training at the National Training Center and the Joint Readiness Training Center, Army ground units conducting training exercises have frequently relied on manned aircraft to replicate the capabilities of the Air Force's Predator and Reaper UAS. Officials told us that the use of manned aircraft in this role permits ground units to practice the process to request and integrate the capabilities provided by Air Force UASs in joint operations. However, this practice is not optimal, as the manned aircraft do not replicate all of the capabilities of the Predator and Reaper aircraft, such as longer dwell times.

GAO, 2010a, p. 26.

Key Areas of Concern

We note above that, as the war in Afghanistan draws to a close, the services must pay close attention not only to the training of UAS crews and support personnel but also to routinely including these systems in combined arms training at home station and at the CTCs. This will require attention to a number of issues.

Beddown and Related Support Facilities

Discussions with officers of the 432d Wing suggest an Air Force future much like the past: flying in support of COCOM requirements worldwide. If this comes about, there will continue to be limited opportunities for the Air Force to support Army or Marine Corps training at the home station or CTCs. Alternatively, if providing such training becomes a priority, the Air Force's current basing and beddown posture will become a problem.

Airspace Considerations

The unique limitations of UASs pertaining to operations in the NAS make integration into home-station training difficult. Currently, UASs can operate only in restricted military airspace or, in the NAS, with COAs from the FAA. These COAs can often be prohibitive to participating in joint training. For example, the MQ-1 FTU from March ARB tried to participate in training with the Marines at Twentynine Palms. Unfortunately, the COA requires a manned chase plane and, given the geography, the contracted chase plane cannot fly the shortest route. The only option is to fly through the restricted airspace between Fort Irwin and Twentynine Palms, which is prohibitively long because the COA permits operations only during daylight hours. These limitations restrict where UASs can be used because an installation must have access to a significant amount of restricted airspace. Other limitations include the basing expense of sprinkling UASs across the country to embed with Army or Marine units at home station.

Several different possibilities could be considered to help increase joint training opportunities. The Air Force could consider the location of Army hubs when choosing where to base its UAS fleet. A location that is near Army maneuver elements might make airspace access less of a restriction or at least make COAs more practical. Proximity might provide more opportunities for training UAS integration throughout the entire mission planning process.

Alternatively, the Air Force could permanently leave a small number of UASs at a training location and remotely operate these from Holloman AFB, Creech AFB, or other locations. These systems are designed to employ remote split operations, in which the crew operating the platform need not be colocated with the platform. The various Air Force units could get flying time by cycling through these dedicated aircraft. This option would provide Air Force crews—including launch and recovery crews—practice integrating with ground forces and would simultaneously support the ground forces' need for integrated training. This plan would also reduce the issue of airspace access because the aircraft would be located within the restricted airspace in which they would be used or at an auxiliary airfield close enough that a COA would be practical.

A promising development that enhances UAS training is the advancement of ground-based sense-and-avoid (GBSAA) and airborne sense-and-avoid (ABSAA) technologies. These may open up regions of civil airspace for properly equipped UASs to operate safely in accordance with the FAA's mandate to "do no harm" without requiring FAA issuance of COAs. In particular, GBSAA has recently completed a series of successful demonstrations, and the

Army is planning to GBSAA equip a number of its UAS training bases to extend their current military airspace with adjoining civil airspace to increase UAS training capacity by 2015. The Army is the lead for the development of GBSAA, but GBSAA has been designed for the use of all the services. Each service, like the Army, would use one of its existing ground based radars in implementing GBSAA. ABSAA's implementation uses UAS-borne sensors and electronics, including those not currently on UASs. Current development of all-weather ABSAA that could operate without FAA COAs is restricted to Global Hawk, with plans to scale down to Reaper. ABSAA availability is not known at this time but will be some years after GBSAA becomes available.

Navy

Relative to the Air Force and Army, the Navy has had limited experience with UASs. To date, Navy ships have employed the Scan Eagle, a non–program of record UAS, fielded through a rapid-acquisition program (see Chapter Two). Scan Eagles are flown from frigates. The Navy has also operated BAMS-D,[16] which is scheduled to be replaced by the MQ-4C, Triton.[17] The Navy is acquiring the Fire Scout (MQ-8B). While we limited most of our study effort to the period through FY 2012, the Navy's most recent experience is the best example of the evolution of UASs and their implications for operational capabilities. In early May 2013, the Navy launched an X-47B experimental drone from the nuclear powered aircraft carrier USS *George H.W. Bush* as it operated off the coast of Virginia (Vergakis, 2013). This launch represented the latest step in introducing a disruptive technology into the operating forces and helped mark a paradigm shift in warfare because this particular UAS is capable of autonomous flight.

The Navy's approaches to integrating UASs and to training and developing UAS operators differ significantly from those of the Army and Air Force. Generally, Navy UASs are being incorporated into existing fleet aviation units and organizations, and the personnel who will operate them are, to a large extent, the same personnel who will operate traditional aircraft assigned to these organizations. In this way, the Navy is addressing several problems often seen when disruptive technologies are introduced. The Navy approach has emphasized the complementarity of the systems, rather than pitting the new technology and its proponents against the old technology and its proponents. In addition, the Navy will reduce cost by capitalizing

[16] While the Air Force has conducted Global Hawk UAS operations since 2001, the Navy has just recently pursued using the system. The Navy purchased Global Hawk airframes from the Air Force and is using them to test operational concepts and technologies. The aim for the Global Hawk maritime demonstration program, BAMS-D, is to better understand maritime surveillance with UASs that fly at high altitude. The goal is to improve the fleet's operational understanding of the battlespace, and the Navy deployed BAMS-D in 2009.

[17] The Navy's BAMS UAS (MQ-4C), now called Triton, is a persistent maritime intelligence, surveillance, and reconnaissance system. As a persistent airborne asset, Triton's sensors will support increased situational awareness for naval forces. The imagery and data provided by the airframe's sensors will be directly available within federated networks, allowing Navy, joint, allied, and coalition exploitation centers to use MQ-4C UAS data.

While the Navy is currently operating only BAMS-D (in FY 2012), it plans to purchase a total of 68 Triton MQ-4C UAS airframes from FY 2014 through FY 2026. This acquisition plan will increase deployment of BAMS to operational commanders throughout the world. The current concept of operations includes plans for the Navy to operate Triton UASs in five separate geographic locations. These operations would provide continuous maritime surveillance, with persistent ISR support 24 hours a day, seven days a week, out to ranges of 2,000 nautical miles from the deployed location. (Derived from Department of the Navy, 2010.)

on investments already made in personnel aviation training. For example, the Fire Scout UAS will be embedded in Navy H-60 squadrons.[18] This enables taking advantage of H-60 pilots' training and experience, because they will also pilot the Fire Scout. The flying skills of the H-60 pilot are utilized for both airframes and minimize the potentially disruptive effects of this new technology. Similarly, Triton, the follow-on to BAMS-D, is intended to complement Navy P-3/8 aircraft with its long loitering capability and sensors. The Navy plans to have P-3/8 officers both be in tactical command of the UAS and pilot the Triton UAS. These officers will first serve and qualify in the P-3/8 aircraft, rotate to a UAS training unit, then assume duties in a UAS squadron. The Triton UAS capabilities complement the capabilities of the P-8 aircraft. The training construct the Navy is pursuing will capitalize, complement, and exploit the capabilities of both aircraft and the personnel assigned to them.

Training for Operators and Support Personnel

Scan Eagle

Qualification training for personnel directly associated with UAS operations is not yet well established for the Navy. Currently, contractors operate the Scan Eagle UAS. (See Figure 4.8.)

Fire Scout

The Navy will have two approaches for Fire Scout operator training—one for littoral combat ship (LCS) employment and another for SOF support. The majority of Fire Scouts will be flown from the LCS. (See Figure 4.8.)

Littoral Combat Ship Employment

The aviation detachment that will deploy with the LCS will consist of one H-60R helicopter and two Fire Scouts and will have 23 assigned personnel. The training these personnel will need is similar to that for deploying with H-60 detachments on guided missile destroyers today (normally, two H-60 helicopters deploy with a guided missile destroyer). In the future, ships will deploy with a mix of H-60s, and Fire Scout and aviation detachment personnel will fly both aircraft depending on mission requirements.

The skills needed to operate and maintain the Fire Scout are complementary to those needed for the H-60. The operators and maintainers of Fire Scout will be the same as for the H-60, with some additional training on the Fire Scout. An H-60 fleet replacement squadron's (FRS's) Fire Scout fleet introduction team will provide this training to the AVOs and MPOs en route. The AVO course for H-60 pilots lasts five weeks. This initial and all sustainment training for the Fire Scout will be conducted via simulator.

Fire Scout Support for Special Operations Forces

The Navy is establishing dedicated units to support SOF, rather than integrating Fire Scout into existing units. The first of these units, Unmanned Helicopter Reconnaissance Squadron One, will be both a training and an operational squadron. The Navy's plan is to organize each such squadron to have nine detachments, each consisting of three Fire Scout airframes, seven to eight AVOs, eight MPOs, and 16 maintenance and support personnel. The Fire Scouts that will be flown off LCSs will be piloted by commissioned officers. The Fire Scouts supporting

---

[18] The Fire Scout will greatly contribute to increased mission readiness and response by conducting persistent ISR operations immediately around seaborne task forces, as well as over the horizon. The persistence, range, sensors, and data sharing of UASs extend the ISR reach of Navy ships and increase situational awareness.

Figure 4.8
Scan Eagle (top) and Fire Scout UASs

SOURCE: DoD.

SOURCE: DoD.
RAND *RR440-4.8*

SOF units will be piloted by enlisted AVOs. Initially, these AVOs will be air warfare operators who have completed tours as MPOs with an LCS Fire Scout.

Fire Scout Maintainers

The Fire Scout maintainers consist of four Navy specialties. Enlisted aviation machinist mates and aviation structural mechanics will take the mechanical course for LCS Fire Scout. Enlisted aviation electronics technicians and aviation electrician's mates will take an electronic technical course for LCS Fire Scout. The maintenance and support skills needed for the Fire Scout are complementary skills for Navy enlisted technicians.

Triton

Operator Training

The MQ-4C Triton UAS is composed of several systems: the airframe; a suite of mission payloads; communications systems; a mission control system (MCS) used for mission planning, control, and execution; and a support system. The Triton UAS will be deployed as an adjunct system to P-3/8 squadrons. The MCS will be based at a main operating base in CONUS, and BAMS maintenance and launch and recovery operations will take place at a forward location.

The Triton watch organization will consist of a mission commander (P-8 naval flight officer),[19] one AVO (officer, P-8 pilot), and two mission payload (enlisted) air warfare operators. Training plans for the Triton watch organization are being developed, and the initial operating capability of Triton will be in FY 2016. The Navy plans to use P-3/8 pilots as tactical coordinators and AVOs on the Triton. Future aviators will first go through flight school, earn wings on the P-3/8 aircraft, then perform an initial operational tour on the P-3/8.[20] After that, aviators will rotate to the FRS for training on Triton before reporting to their squadrons. Designated FRS for BAMS is Patrol Squadron 30 (VP-30), located at Naval Air Station Jacksonville, Florida. FRS training provides initial and refresher qualification training for personnel who will be assigned to Triton units.

Maintenance personnel and the pilot who launches and recovers the airframe are located at the forward operating base. The Navy plans to rotate personnel through the forward base every several months.

A simulation training capability will be built into the MCS. Initial Navy manning plans indicate that there will be eight to ten crews per squadron. The crews will be flying real missions that will sustain their proficiency. Triton UASs are flown via point and click; there is no stick and rudder. VP-30 is developing the tailored Naval Air Training and Operating Procedures Standardization requirements for the Triton; the final training and readiness manual for Triton is due in 2013.

Maintenance Training

The Triton airframe, engines, and associated equipment are different from the P-3/8; one is a UAS and the other a manned aircraft. Separate training will thus be necessary. When the

---

[19] The mission control officer is a naval flight officer who is responsible for the mission planning and the overall tactical employment of the airframe. The AVO is a naval aviator, responsible for flight planning and safety of flight, and is the pilot in command of the airframe. The MPOs are enlisted air warfare operators, responsible for the operation and employment of the sensors and the detection and analysis of targets.

[20] The notional pilot training track for P-3/8 pilots is (1) attend and graduate from flight school, (2) attend FRS for P-8 training, and (3) conduct an operational P-8 squadron tour.

Triton UAS is introduced into the fleet, the Chief of Naval Air Technical Training will stand up a technical school to train enlisted personnel to perform Triton maintenance.

Operational Training

BAMS/Triton

The P-8A's missions overlap with those of Triton. The missions the two have in common include maintaining the maritime common operational picture and the classification, identification, detection, and tracking of surface units. The P-8A, however, has missions that the Triton does not perform (e.g., antisubmarine warfare) and vice versa. However, the increased utilization and cross training of P-8A and Triton crews over time can and is expected to increase understanding and employment of these systems in support of the warfighter.

While the Navy was operating only BAMS-D in FY 2012,[21] the Navy has plans to purchase a total of 68 Triton MQ-4Cs UAS airframes from FY 2014 through FY 2026. This acquisition plan will increase deployment of Triton UAS to operational commanders throughout the world. The current concept of operations includes plans for the Navy to operate Triton UAS in five geographic locations. These operations would provide continuous maritime surveillance, with persistent ISR support 24 hours a day, seven days a week, out to ranges of 2,000 nautical miles from the deployed location (Department of the Navy, 2010). The LCS does not have the Navy Continuous Training Environment (NCTE) synthetic training capability, and Fire Scouts assigned to the LCS will have no capability to train in the NCTE. Integrated training plans for Fire Scout are being developed.

Integrated Training

Triton's MCS will be configured so that it can participate in fleet training exercises synthetically via NCTE. Fleet synthetic training events for BAMS-D and Triton are being developed.

Key Areas of Concern

Airspace Considerations

Airspace planning is a critical requirement when conducting UAS operations but is less of a constraint for UASs in naval operations. The FAA controls the airspace up to the 12-mile limit from land. Navy surface ships with Scan Eagle and Fire Scout can embark the UASs, travel 12 nautical miles out to sea to a naval operating area, then conduct UAS training.

BAMS-D and Triton use special use airspace and fly at altitudes higher than commercial airlines do. They file instrument flight rules flight plans. However, BAMS-D and Triton do not incorporate sense-and-avoid technology, and conflicts can exist with visual flight rules aircraft in the airspace. Airspace issues exist at some Navy training installations as well. For example, along the North Carolina coast, the expeditionary readiness group must work with the FAA to route commercial aircraft around the established restricted operating zone units.

---

[21] BAMS-D has participated in CSG training events in the Virginia Capes operating area, including combined training unit exercises and joint task force exercises. Navy officials emphasized that Triton is a tactical asset and will respond to the requirements of the task force.

The Need for Interoperability

So far, this report has emphasized that home-station training is not as effective as it should be, thus reducing the effectiveness of multiservice, or interoperability, training at CTCs and during joint exercises. The importance of interoperability training is stressed in Chairman of the Joint Chiefs of Staff Guide 3501 (2012, pp. B-2 and B-3), which notes that the

> ability of systems, units, or forces to operate in synergy in the execution of assigned tasks is critical to successful operations. This ability to operate effectively together describes interoperability. From a joint training perspective, interoperability is a Service component responsibility. Interoperability training is based on joint doctrine, or where no joint doctrine exists, on Service or [SOF] doctrine to prepare forces or staffs from more than one Service component to respond to operational and tactical requirements deemed necessary by CCDRs [combatant commanders] to execute their assigned missions. Interoperability training involves forces of two or more Service components (including SOF) with no interaction with a CCDR or subordinate JFC [joint force commander] or joint staff.

The publication in 2011 of the MTTP for UASs was designed to address the problem of interoperability. As designed, it is

> a single source, descriptive reference guide to ensure effective planning, integration, and utilization of multi-service UAS capability. It provides commanders, operational staffs, requestors, and UAS operators with a comprehensive resource for planning and employing unmanned aircraft . . . . US Air Force (USAF) uses the term [RPA] for the air vehicle component of a UAS. (MTTP, 2011, p. i)

To date, interoperability training at the NTC and the Joint Readiness Training Center has been limited.[22] GAO (2005, p. 2) cited two significant challenges for improving interoperability training: (1) "establishing effective partnerships with program stakeholders through comprehensive communication and coordination and (2) developing joint training requirements that meet combatant commanders' needs," with particular emphasis on tactical level training (GAO, 2005, p. 2).[23] To meet this challenge, DoD established the Training Transformation Implementation Plan (Director, Readiness and Training Policy and Programs, 2006), giving the Office of the Under Secretary of Defense for Personnel and Readiness overall responsibility and giving the Deputy Under Secretary of Defense for Readiness executive agent responsibility for training transformation planning, programming, budgeting, and execution progress.

---

[22] GAO, 2005, p. 1, found that

> U.S. forces are conducting significantly more complex operations, requiring increased interoperability between and among the military services, combatant commands, and other DOD and non-DOD organizations. In the past, military services experienced some joint operations training during joint exercises, but most service training focused on individual service competencies with limited joint context.

[23] GAO, 2005, p. 18, noted that,

> in the past, joint training tasks were primarily focused at the command level and were identified through DOD authoritative processes that built requirements by translating command combat commanders inputs into training requirements. Training transformation has expanded joint training requirements to include those at the tactical level in addition to joint command level training.

While RAND did not assess interoperability training in detail, discussions with the Army's training community, as noted above and as reported in GAO, 2010a, p. 26, suggest that such opportunities have been limited by the lack of Air Force assets, particularly the Predator, due to pressing operational requirements in Iraq and Afghanistan. The problem of transiting Air Force UAVs from Creech AFB to the NTC at Fort Irwin through FAA controlled airspace is often cited. Permanently stationing Air Force UAVs at the NTC would, of course, negate that problem, much as the Marine Corps has done by basing UASs at the MCAGCC at Twentynine Palms, California, to support training there. However, the experience of the Marine Corps suggests that the mere presence of UASs will not ensure that the forces are properly trained. The forces must prepare at their home stations for such training. In addition, given the common MTTP, forces trained to operate with their own services' UASs should find working with UASs from the other services less challenging.

Simulators

Congress has pressed for information on the role that simulators might play in a UAS training strategy, seeking an informed balance between live training and simulated training. In addition, the GAO reported that the

> military services lacked simulators that were capable of supporting training that is intended to build proficiency in skills required of UAS vehicle and sensor operators and prepare these personnel to conduct UAS combat missions. ... [And] the Air Force and the Army have not fully developed comprehensive plans that address long-term UAS simulator requirements and associated funding needs. (GAO, 2010a, pp. 27–28)

Implicit in the GAO's comments is an assumption that simulators should play an important role in the future of UAS training. Our review questions that conclusion.

DoD has historically viewed simulator training and joint use of training facilities as an exercise in cost reduction. It is first necessary to understand how simulation better enables the introduction of the technology and, similarly, how joint use of the disruptive technology enables more effective military operations. For the purposes of this report, we have clearly focused on these aspects of simulation and jointness, which must be understood before the more-traditional perspectives of resource allocation can be applied. Currently, the usual significant savings that the department has garnered from simulators in the past do not appear to obtain in this instance. The danger is that, if the department starts with a savings orientation, it will not fully capitalize on the capabilities UASs represent.

For traditional aircraft systems, one major appeal of using simulators, rather than live flying, is the cost differential. For example, a RAND study of the trade-off between live and simulated training for manned aircraft noted that "the Marines and British are starting to give greater recognition to use of simulators as they seek either to reduce the high costs of live training or conserve the limited life of operational aircraft" (Schank et al., 2002, p. 49). In this case, the cost of flying the strike fighter aircraft they were discussing was many times greater than the cost of using a simulator, and the procurement cost of the aircraft itself was also significantly greater than that of the simulator. That said, they also noted that,

> [t]o increase that use, several improvements must happen to integrate simulators more fully into their unit training. The fidelity and availability of simulators must increase to the point that fighter pilots see their benefit in training. This requires additional funding for simulators. (Schank et al., 2002, p. 49)

The report further argues that, even given the cost advantage of simulators for strike fighter aircraft training, simulators are a complement to live training, not a substitute for it.

The same point was made to us concerning UAS simulators. Officials of the 432d Wing also noted that simulators were not very good for training on landings. Moreover, a substantial investment in new simulator technology would be necessary to improve the realism of UAS simulators to the point it could prove useful for training UAS crews and ground forces. For example, GAO, 2010a, p. 28, found that currently Air Force and Army simulators did not have the ability

> to replicate all UAS procedures and to enable the integration of UAS training with other types of aircraft, [e.g.,] ... the Army's Shadow Institutional Mission Simulator is not currently capable of replicating system upgrades that are being fielded directly to ongoing combat operations, such as a laser target designator and communications relay equipment. ... Air Force and Army simulators are also ... incapable of providing virtual, integrated training opportunities between manned and unmanned aircraft because of interoperability and information security concerns.

The conditions that make a compelling case for simulators for the training of fighter pilots are largely absent when it comes to UAS training:

- First, UAS flying hours are much less expensive than flying hours for manned aircraft. By one account, the flying hour cost of Shadow was $705; a comparable figure for Gray Eagle was $4,275.[24] By comparison, the Air Force told the GAO that "the live training cost of one F-15E flight hour is approximately $17,449" (Pickup, 2012, pp. 15–16). Other sources peg the flying hour costs of manned aircraft much higher.[25]
- Second, the kinds of operational activities that need more training emphasis, in particular, air-ground coordination, are generally not activities in which simulators *per se* can substitute for live flying. A good example of both the limitations and the value of simulators can be found at the digital air-ground integration ranges the Army operates today. On these ranges, only the weapon system engagement is simulated. Everything else—planning, coordination, UAS operation, ground unit and/or manned aircraft activity, target acquisition—is live.
- Third, while simulators are inherent in the systems used for initial training of pilots and sensor operators,[26] they are not well suited (and not designed) for training operating

---

[24] Ron Moring, UAS cost per flying hours, personal communication, April 5, 2013. The cost for the Gray Eagle is similar to the costs the Air Force reported for Predator at $3,679 and Reaper at $4,762, as reported by Mark Thompson, "Costly Flight Hours," *Time*, April 2, 2013.

[25] Recently, the Air Force comptroller's office told *Time* that the flying hour cost for the F-15C was $41,921. Other aircraft had considerably higher flying hour costs; e.g., the cost of flying an F-22 for one hour was reported to be $68,362. See Thompson, 2013.

[26] See Shawn Johnson, DAMO-TRC Army Flying Hour Program, "UAS cost per flying hours," personal communication, with Joan Vandervort, Office of the Office USDP&R, April 10, 2013.

forces on the full spectrum of UAS capabilities. Simulators may be useful as a supplement to live training of pilots and crews but do not replicate the myriad activities and stimuli that a live training environment provides well for maneuver forces.

- Fourth, as noted previously, simulators at best complement rather than substitute for live training. Given the current state of UAS fielding, first priority must be given to building the live-flying infrastructure. Then, and only then, might funds be used to undertake the research and development that must precede any consideration of fielding the kind of interactive UAS air-ground simulators that could train both UAS crews and ground troops.

In the future, the services should reevaluate the need for virtual training using a cost and effectiveness analysis (COEA) including developmental costs for simulators, the savings based on the costs of flying UASs, and the critical availability of air space for training. Given current budget limitations and the importance of fully developing the opportunities for live training and the relatively low cost of such training, diversion of funds to a research and development program to develop higher-fidelity UAS-ground simulators is not appropriate at this time.

Summary

This chapter presented an overview of UAS training in 2012. Our focus was on service-specific and interoperability training. The training framework presented in Chapter Three was our starting point, with an emphasis on preparing forces for joint training and deployment. We looked at ground, air, and sea forces. Insights came from a wide variety of sources but mainly through visits to selected service bases, where we talked with operators and support personnel, those who train operators and associated team members, and those who train and observe the forces who employ UASs. In general, the RAND team found that qualification training in all the services for designated UAS-specific MOSs is well established. The critical issue is being able to train with ground forces at home station. The experience of the Marine Corps illustrates the importance of such opportunities. The Marine Corps has Shadow squadrons based at the MCAGCC at Twentynine Palms, California, where the restricted airspace is adequate both for maintaining crew proficiency and for supporting ground forces. However, both the ground trainers and squadron personnel the RAND team interviewed reported that the ground forces are not adequately trained at home station to integrate UAS capabilities into their operations. This required starting with what amounted to "remedial" training at Twentynine Palms, and trainees thus did not achieve the level of proficiency desired. This suggests that the biggest challenge in developing and implementing a training strategy for UASs is gaining and maintaining proficiency on the part of end users or "supported units"—the leaders and staffs who will integrate UAS capabilities into their operations.

Thus, as the war in Afghanistan draws to a close, the services must pay close attention not only to training UAS crews and support personnel but also to routinely exercising these systems during combined arms training at home station and at the CTCs. We addressed a number of issues for this, including beddown and support facilities and airspace considerations.

We found that home station training is generally not as effective as it should be, thus reducing the potential effectiveness of multiservice or "interoperability" training at CTCs and during joint exercises. We noted that, to date, interoperability training at the CTCs has been limited by the lack of Air Force assets, particularly the Predator, due to pressing operational requirements in Iraq and Afghanistan. However, given the common MTTP, forces trained to operate with their own service UASs should find working with UASs from the other services less challenging.

Finally, we considered the case for UAS simulators. Currently, such simulators do not appear to offer the significant savings that the Department has usually garnered from simulation in the past. We concluded that, given current budget limitations and the importance of fully developing the opportunities for live training and the relatively low cost of such training, diversion of funds to a research and development program to develop higher fidelity UAS-ground simulators would be unwise at this time.

CHAPTER FIVE

Implications and Recommendations

The Guiding Principle: "Train As We Fight"

DoD has been under some pressure from Congress and such organizations as GAO for the lack of central control and the lack of the development of a joint DoD strategy for training. For example, "DOD Lacks a Comprehensive, Results-Oriented Strategy to Resolve UAS Training Challenges," was the summary title of a section of a recent GAO report (GAO, 2010a, p. 29). The Joint Unmanned Aircraft System Center of Excellence was established to support the joint operator and the services by facilitating the development and integration of common unmanned aircraft system operating standards, capabilities, concepts, technologies, doctrine, tactics, techniques, procedures, and training. The center, however, was disestablished in May 2011, and its UAS oversight function was reassigned to the Joint Staff Force Structure, Resources, and Assessment Directorate (J8), with some responsibilities assigned to the UAS Task Force. However, the 2011 publication of the MTTP for UASs addressed the problem of interoperability, and the services have agreed to incorporate the MTTP in their training. While the services have agreed to a common set of TTP, each service continues to develop its own UASs to meet its own requirements. Moreover, the GAO (2010a, p. 25) has well documented that the opportunity for joint training is limited. While the GAO identified a number of organizations and initiatives addressing UAS training challenges, only one of the eight organizations and initiatives it listed were led by the services. Given the current maturation of UAS platforms, this emphasis is misplaced; the military services employs their UASs in very different ways. Today, the services are still struggling with how to incorporate these new systems. The appropriate strategy for UAS training is to encourage each service to solve its own UAS training problems, then to coordinate the resulting TTPs by updating the MTTP to achieve a high level of interoperability.

Commanders and operators from the bottom up are discovering and adapting to the revolutionary operational changes that UASs have brought about. The most apparent changes are in the persistence and responsiveness of UASs relative to other platforms in the permissive threat environments in which they are currently operating. These characteristics help develop valuable shared situational awareness between UASs and the supported forces and also make ISR capabilities more responsive and immediately available. Processing, exploitation, and dissemination concepts for UAS-generated information, however, are still evolving. In addition, armed UASs provide strike capabilities similar in some ways to those of CAS; the evolution of UASs from pure ISR platforms to small, lightly armed ISR platforms, to bigger, more heavily armed, faster, and even stealthier reconnaissance-strike platforms has crossed operational and cultural seams. This has lead to the need for continual resolution of disputes between intelligence and operations staffs over priorities for UASs and their implications for mission planning and execution.

To date, the services have, in a variety of important ways, been "training *while* they fight." Iraq and Afghanistan have served as de facto training ranges, as well as an active theater of war. For example, Air Force Predator and Reaper pilots and sensor operators have frequently gone straight from flight instruction and certification to flying real-world missions, learning and adapting as they support operations. Army and Marine Corps units have also learned in theater how to capitalize on the robust set of capabilities UASs provide.

The demonstrated ability of the services to learn on the fly should be cause for optimism about continued success in UAS training as operations wind down. But that continued success will depend on preserving training opportunities in an environment in which training is not an incidental by-product of operational requirements. Operational experiences, for example, typically do not suffer from the inherent limitations of CONUS training venues, such as limited area, limited time, and limited fidelity. Because of such limitations, it has been harder to train as we fight in CONUS. Infrastructure and other limitations currently make it difficult or impossible to train at night or in bad weather, to train with live or dummy munitions, or to train in a communications environment that represents the environment and demands of actual operations in theater. Preservation of the current levels of proficiency, improvement where needed, and adaptation to changing circumstances will not come without effort; they will require overcoming or at least alleviating the limitations mentioned here.

Elements of a Strategic Training Plan

To *train as we fight*, the DoD training strategy must

- engender better appreciation of UAS capabilities throughout the chain of command
- address organizational, structural, and infrastructure and support issues
- enable training as we fight in collective unit training
- enable training as we fight in larger exercises.

Engender Better Appreciation of UAS Capabilities

Many ground force commanders and their battle staffs today do not fully recognize the potential benefits UASs bring to the fight and, as a result, are not in a position to benefit from them in circumstances that would improve units' warfighting capabilities. This is traceable in part to limited use of UASs in home-station and CTC training but also to other inherent limitations, such as a shortage of experienced personnel and UAS assets and airspace and infrastructure limitations. Overcoming these limitations will enable more opportunities to foster better understanding and improve integration. Training guidance should emphasize the value of UASs, using examples of how units have benefited by preparing themselves to use such systems through training and attending familiarization courses. Such vignettes can be inserted into professional education and training materials to share positive experiences working with UASs.

Address Organizational, Structural, and Training Infrastructure and Support Issues

Organization and Structure

UASs are a disruptive technology, and their introduction can create tensions between operations and intelligence staffs on the right balance between using UASs to collect intelligence or launch strikes. In the Army, for example, one could ask, is Gray Eagle an ISR asset or an armed reconnaissance system? Doctrine development efforts should produce a doctrine flexible enough to allow changes in allocation and organizational balances and address the attendant training issues accordingly.

Also, in the Army, the organizational placement of UASs remains a challenge, as illustrated by the fact that the assignment of responsibilities for the Raven is still being worked out. A more-systematic approach to monitoring and maintaining Raven operator qualifications will better enable integration of the system into maneuver forces and thus improve overall training and help develop better appreciation of the system's capabilities and limitations. In addition, the Army is currently examining alternatives for placing Shadow platoons in the headquarters company of a brigade combat team or consolidating them in the developing full-spectrum combat aviation brigade. Resolution of these organizational issues will likewise aid in training integration.

Training Infrastructure, Training Support Issues, and Related Factors

Some of the most important ways in which shortfalls can detract from UAS training are

- incompatible or different communications capabilities
- airspace restrictions
- inability to train at night or in bad weather
- inability to use live or dummy munitions
- UAS unavailability
- insufficient time.

These shortfalls cannot always be fully corrected. Some, such as differences in communications capabilities, are inherent but can be alleviated to some extent. The full correction of others may be unaffordable, but partial correction should still be possible. Inability to train at night or in bad weather, for example, is to some extent a result of FAA restrictions on when Air Force UASs can fly from their bases to the NTC. Basing Air Force UASs at the NTC would significantly reduce the effect of these restrictions and might also make it possible for the UASs to use live or dummy munitions, further improving realism for both customers and operators.

Research suggests that airspace restrictions, while sometimes nettlesome, need not pose a serious obstacle to most UAS training at a unit's home station, either for operators or end users. Nevertheless, efforts to gain more access to airspace should continue with the research and development of the joint program for GBSAA systems, led by the Army's program manager for UAS. This system uses ground-based radars to provide an extra layer of airspace safety, which could enable the FAA to make the use of civil airspace adjacent to UAS training airspace less restrictive and thus expand the airspace available for UAS training. A promising initial demonstration with Gray Eagle took place in summer 2011, but GBSAA is not expected to be certified for use until 2015.

The insufficient number of UASs for training should be alleviated to some extent as some systems redeploy from overseas theaters and also by the recent decision to increase the number of Predator/Reaper CAPs from 61 available today to 85 in the future. Similarly, the Operation Enduring Freedom drawdown will likely free up more RPA trainers, alleviating the shortfall.

Enable Training as We Fight in Unit Collective Training

To facilitate home station training, the following steps need to be taken:

- Training must make full use of units' UASs and must use simulation or surrogates, despite their limitations, when live use is not possible.
- Training should include the systems of other services and their associated joint tactical air controllers (JTACs) and air liaison officers (ALOs) to the extent possible.
- Air Force preparation training of JTACs and ALOs should be available for support of unit-level training, as well as for larger exercises.
- After-action reviews and reports should be expanded to reinforce lessons learned; they can also provide a sound basis for continued measurable improvements. Over time, the reports growing out of these documents should become more detailed and should incorporate measures of performance and effectiveness, so that these metrics can drive future improvements.

Enable Training as We Fight in Larger Exercises

By design, larger exercises like those at maneuver CTCs provide considerable opportunities for more-realistic training. They are thus already better postured to provide training on the use and integration of UAS. However, more can be done to enhance the use of training opportunities in such large exercises to further the goal of better UAS integration in the force. Specifically, we recommend

- continued stress on training realism
- including live UASs and their associated JTACs and ALOs as much as possible
- including the CAS-like and ISR missions that UASs, especially Air Force UASs, will be performing in theater
- providing a small paved air strip to base RPAs, launch and recovery elements (LREs), and support crews at the maneuver CTCs to avoid the NAS constraints associated with remote operation
- continued efforts to integrate additional Creech AFB–based UASs into CTC training, when needed.

Initiatives with Likely Near-Term Payoffs

Other things can be done that should provide near-term payoffs:

- **Increase exposure to the capabilities and limitations of UAS**—Continue to expose appropriate commanders to UAS operations through short (one- to two-day) familiarization courses, such as those the Air Force offers at various UAS squadrons' locations (e.g., Creech AFB, Beale AFB, and March ARB). These also need to be introduced as part of the joint training curriculum in a more formal and standardized fashion across the force. The proliferation of Air Force UAS squadrons geographically within CONUS should provide ready access for ground units whatever their home station.

- **Harness the lessons-learned process to guide the development of service and joint doctrine and TTP**—Formalize and standardize these processes as appropriate, and share best practices more extensively across services and regions. Right now, most extended learning is done "on the job," with lessons learned shared informally among operators and trainers. Different UAS elements tend to focus support on different ground units, each with geographical and other mission differences. Best practices across these geographic and mission boundaries need to be understood to inform joint doctrine and, hence, responsive training requirements.
- **Address well-known but underresourced training infrastructure shortfalls**—Consider this issue in light of our previous discussions of basing. UASs can be based at maneuver CTCs to minimize the impact of FAA restrictions. If this is done as a part of the drawdown and redeployment of RPA units to CONUS, this should not be too disruptive or expensive; many details remain to be worked out, but military construction and other costs appear to be manageable. Communications infrastructure and C2 organizations and processes at the ranges can be enhanced to better mirror operational capabilities.

Institutionalizing Training for UAS Capabilities over the Longer Term

Institutionalizing UAS training over the longer term will naturally depend on continued support for the specific programs and initiatives and others like them that we outlined earlier. Institutionalization will also require more-general efforts, falling into three broad categories: acculturation, professional development of the UAS force (operators, maintainers, and a UAS chain of command), and integration of the UAS community (a subprofession, if you will) into the aviation profession. These three are somewhat interwoven, and the third depends on the first two. Finally, institutionalization over the longer term will require adaptation: The strategies must be designed to be adaptive, and those who develop and implement them must be willing to adapt them.

Acculturation

The RAND team heard from many sources that many end users do not yet have a well-developed appreciation of the capabilities UASs can bring to an operation. In some cases, this lack of appreciation has derived from negative experiences. Some of these experiences resulted from inexperienced operators, some from mechanical failures, some from sheer bad luck (e.g., weather). The need to deconflict airspace and the need to do so, even when using restricted airspace reserved for military use, often led to utilization problems and contributed to negative impressions about the value of UAS. First impressions do count, and avoiding bad ones is much easier than overcoming them later.

Professional Development

All the services need to maintain professional development strategy for the UAS force. The services are developing a professional force of operators for UASs, both systems and payloads, consisting of enlisted soldiers, noncommissioned officers, and warrant and commissioned officers, although the Army currently has no UAS officer specialty. It is important to develop a cadre of UAS professionals with hands-on experience in all echelons of UAS operation; the Army may wish to consider a career pipeline for such officers, perhaps with an additional skill identifier. Such a professional cadre is a key element in full integration of UAS capabilities.

Integration

The services are making significant efforts to integrate UASs into their force structures. Army efforts would be aided by more access during training to the operational and even strategic capabilities of Air Force UASs. These efforts will be aided by better acculturation and an institutionalized UAS career field that, over time, develops professionals with a more-complete understanding of UAS capabilities and limitations and sufficient rank, technical expertise, and operational experience to enable better integration of the capabilities into the overall operations of the end users. Related issues include integration of UAS capabilities with other ISR systems and with the intelligence warfighting function more generally, the tension between those who value the strike capabilities of UASs and those who value the intelligence collection capabilities, and the further tensions likely to emerge as UAS capabilities widen further. As UAS capabilities grow and broaden, integrating them into other warfighting functions will also present challenges.

Adaptation

The training strategy must adapt to future technical innovations and changes in operational demands. To enable this adaptability, those concerned with overall training strategy development will need to remain cognizant of and to guide the following, where possible:

- evolution of the roles of UASs in the full range of military operations, and the speed and breadth of the expansion of those roles
- surfacing and resolving UAS doctrinal issues (In particular, how will the requirements of intelligence and operations be integrated and deconflicted? How will both organic and nonorganic systems be better integrated to support ground forces, and which warfighting functions will they support? How will technological advances and changes in expected operational conditions influence decisions regarding these roles and missions?)
- service adjustments of force structures and personnel management processes to accommodate the disruptive technology effects of UASs
- influence of the above and related issues on the development, promulgation, and adaptation of joint and service doctrines so these doctrines can continue to form a solid basis for UAS training.

Summary of Findings and Recommendations

Our findings and recommendations are summarized as follows:

1. The proper strategy for DoD at this time is to encourage each service to solve its own UAS training problems, rather than to constrain any one service under the guise of joint operations.
2. DoD should support current and future programs to develop ranges and beddown and support facilities similar to those in the Army's current programs.

3. In spite of the demands of deploying units and sustaining combat operations in Afghanistan, Army trainers indicate that Army units are better at maintaining qualification for Shadow qualification than for Raven. Nevertheless, not all units have been able to maintain high Shadow qualification rates—some units arrive at the NTC with one-half or fewer of their operators qualified. Better tracking of operators and their sustainment training is needed, particularly but not exclusively for Ravens.
4. If joint training becomes a priority, the Air Force's current basing and beddown posture will become more of a problem. The Air Force could consider the location of Army hubs when choosing where to base its UAS fleet. A location that is near Army maneuver elements might make airspace access less of a restriction or at least make COAs more practical. Proximity might provide more opportunities for training UAS integration throughout the entire mission planning process.
5. Given current budget limitations and the importance of fully developing the opportunities for live training and the relatively low cost of such training, diversion of funds to a research and development program to develop higher fidelity simulators would seem unwise at this time.

Path to the Future

We have presented and discussed a multitude of means for setting up UAS training strategies for success. These means include support for ongoing facilities and basing initiatives; expansion of such facilities, where possible, to enable wider availability of UASs to support collective training; and efforts to increase the use of UASs—the complete UAS package, including JTACs and other coordinating elements—in collective training, both local and in larger exercises. We have noted the need to support continuing efforts to resolve airspace access issues, observing as we did so that there are ways to keep such restrictions from imposing serious limitations on much of the UAS training envisioned here. Similarly, we have discussed the potential for simulators to add value in training, now and perhaps more in the future. But we have also cautioned that simulators, in their current state, are not a good substitute for live use of UASs in collective training.

The path to the future for UAS strategies starts now, with support for ongoing initiatives that will continue the trends towards better training integration and thus better ability on the part of end users to employ the multiple capabilities of UASs in their operations. The path continues, using that foundation, with longer-term efforts to add to acculturation of end users, professionalization of the UAS community, and integration of the two to harmonize the capabilities of UASs as key elements of overall force effectiveness.

APPENDIX A

Current Major DoD UAS Programs in FY 2012

The following pages come directly from a Congressional Research Service report on U.S. UASs (Gertler, 2012, pp. 31–46).

Current Major DOD UAS Programs

This section addresses the program status and funding of some of the most prominent UAS programs being pursued by DOD, and most likely to compete for congressional attention. This section does not attempt to provide a comprehensive survey of all UAS programs, nor to develop a classification system for different types of UAS (e.g., operational vs. developmental, single mission vs. multi mission, long range vs. short range). One exception is a short subsection below titled "Small UAVs." The UAVs described in this section are distinguished from the proceeding UAVs by being man-portable and of short range and loiter time. These smaller UAVs are not currently, and are unlikely to be, weaponized. The services do not provide as detailed cost and budget documentation for these UAVs as they do for major UAS programs. Individually, these UAVs appear very popular with ground forces, yet do not necessarily demand as much congressional attention as larger UAS programs like Predator or Global Hawk. As a whole, however, these small, man-portable UAVs appear likely to increasingly compete with major UAS programs for congressional attention and funding.

Table 6. Characteristics of Selected Tactical and Theater-Level Unmanned Aircraft

System	Length (ft)	Wingspan (ft)	Gross weight (lbs)	Payload capacity (lbs)	Endurance (hours)	Maximum altitude (ft)
Predator	27	55	2,250	450	24+	25,000
Grey Eagle	28	56	3,200	800	40	25,000
Reaper	36	66	10,500	3,750	24	50,000
Shadow	11	14	375	60	6	15,000
Fire Scout	23	28	3,150	600	6+	20,000
Global Hawk	48	131	32,250	3,000	28	60,000
BAMS	48	131	32,250	3,200	34+	60,000

Figure 5. U.S. Medium-Sized and Large Unmanned Aircraft Systems

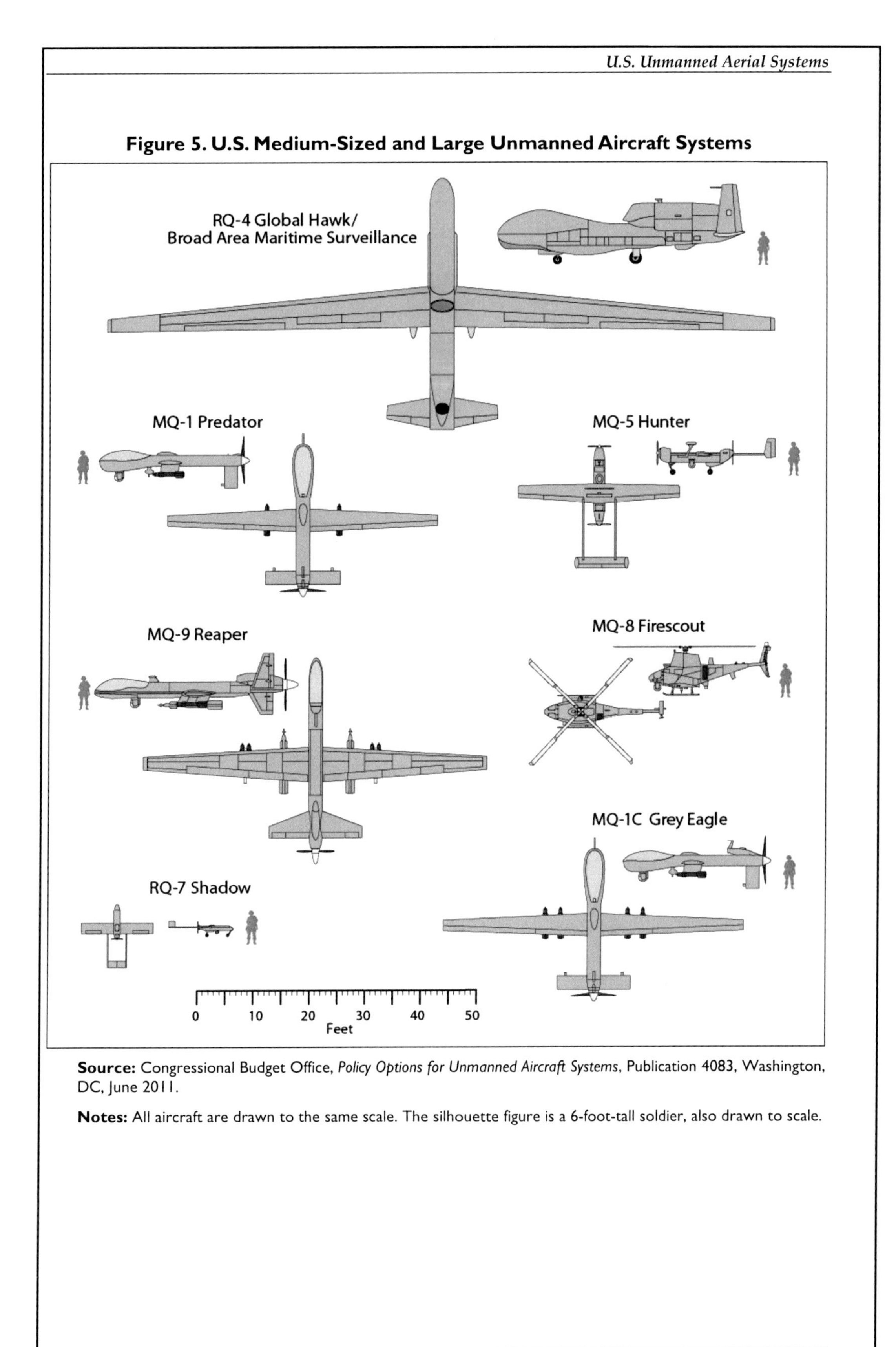

Source: Congressional Budget Office, *Policy Options for Unmanned Aircraft Systems*, Publication 4083, Washington, DC, June 2011.

Notes: All aircraft are drawn to the same scale. The silhouette figure is a 6-foot-tall soldier, also drawn to scale.

Table 7. Acquisition Cost of Medium-Sized and Large Unmanned Aircraft Systems Under the Department of Defense's 2012 Plan

(Millions of 2011 dollars)

	2011	2012	2013	2014	2015	2016	2017	2018	2019	2020	Total, 2011-2020
					Air Force						
RQ-4 Global Hawk	1,200	1,060	890	790	810	710	1,160	530	80	60	7,290
MQ-1 Predator	30	10	10	10	a	a	a	a	a	a	60
MQ-9 Reaper	1,700	1,550	1,740	1,440	1,350	1,150	1,060[b]	1,040[b]	1,030[b]	1,010[b]	13,070
					Army						
MQ-1C Grey Eagle	870	1,060	1,040	740	220	90	a	a	a	a	4,020
RQ-7 Shadow	610	250	270	200	300	280	a	a	a	a	1,910
					Navy and Marine Corps						
RQ-4 Broad Area Maritime Surveillance	530	560	760	880	900	1,010	1,230	1,260	1,130	1,130	9,390
MQ-8 Fire Scout	60	70	60	80	80	90	130	160	150	150	1,030
RQ-7 Shadow	90	10	10	10	a	a	a	a	a	a	120
					All services						
	5,090	4,570	4,780	4,150	3,660	3,330	3,580	2,990	2,390	2,350	36,890

Source: Congressional Budget Office based on data from the Department of Defense's budget request for 2012, Selected Acquisition Reports for December 2010, and Aircraft Procurement Plan: Fiscal Years 2012-2041 (submitted with the 2012 budget, March 2011).

Notes: Acquisition cost includes the cost of procuring air vehicles, sensors, and grounds stations, plus the cost for research. development, test, and evaluation. The services' cost data have been adjusted using CBO's projection of inflation and rounded to the nearest $10 million.

a. The Department of Defense has no plans to acquire or modify the specified system in these years.

b. The cost is for the follow-on aircraft the Air Force plans to acquire instead of the Reaper.

MQ-1 Predator

Through its high-profile use in Iraq and Afghanistan and its multi-mission capabilities, the MQ-1 Predator has become the Department of Defense's most recognizable UAS. Developed by General Atomics Aeronautical Systems in San Diego, CA, the Predator has helped to define the modern role of UAS with its integrated surveillance payload and armament capabilities. Consequently, Predator has enjoyed accelerated development schedules as well as increased procurement funding. The wide employment of the MQ-1 has also facilitated the development of other closely related UAS (described below) designed for a variety of missions.

System Characteristics. Predator is a medium-altitude, long-endurance UAS. At 27 feet long, 7 feet high and with a 48-foot wingspan, it has long, thin wings and a tail like an inverted "V." The Predator typically operates at 10,000 to 15,000 feet to get the best imagery from its video cameras, although it has the ability to reach a maximum altitude of 25,000 feet. Each vehicle can remain on station, over 500 nautical miles away from its base, for 24 hours before returning home. The Air Force's Predator fleet is operated by the 15th and 17th Reconnaissance Squadrons out of Creech Air Force Base, NV; the 11th Reconnaissance Squadron provides training. A second control station has been established at Whiteman AFB, MO.[104] Further, "[t]here are plans to set up Predator operations at bases in Arizona, California, New York, North Dakota, and Texas."[105] The Air Force has about 175 Predators;[106] the CIA reportedly owns and operates several Predators as well.

Mission and Payload. The Predator's primary function is reconnaissance and target acquisition of potential ground targets. To accomplish this mission, the Predator is outfitted with a 450-lb surveillance payload, which includes two electro-optical (E-O) cameras and one infrared (IR) camera for use at night. These cameras are housed in a ball-shaped turret that can be easily seen underneath the vehicle's nose. The Predator is also equipped with a Multi-Spectral Targeting System (MTS) sensor ball which adds a laser designator to the E-O/IR payload that allows the Predator to track moving targets. Additionally, the Predator's payload includes a synthetic aperture radar (SAR), which enables the UAS to "see" through inclement weather. The Predator's satellite communications provide for beyond line-of-sight operations. In 2001, as a secondary function, the Predator was outfitted with the ability to carry two Hellfire missiles. Previously, the Predator identified a target and relayed the coordinates to a manned aircraft, which then engaged the target. The addition of this anti-tank ordnance enables the UAS to launch a precision attack on a time sensitive target with a minimized "sensor-to-shoot" time cycle. Consequently, the Air Force changed the Predator's military designation from RQ-1B (reconnaissance unmanned) to the MQ-1 (multi-mission unmanned).[107] The air vehicle launches and lands like a regular aircraft, but is controlled by a pilot on the ground using a joystick.

MQ-1C Grey Eagle

A slightly larger, longer-endurance version of the Predator, the Army's MQ-1C Grey Eagle entered low-rate initial production on March 29, 2010.[108] The Grey Eagle can remain aloft for 36 hours, 12 hours longer than its Air Force sibling.

An Army platoon operates four aircraft with electro-optical/infrared and/or laser rangefinder/designator payloads, communications relay equipment, and up to four Hellfire

[104] Phillip O'Connor, "Drones seeking terrorists guided from Missouri air base," *St. Louis Post-Dispatch*, May 10, 2011.

[105] P.W. Singer, *Wired for War* (New York: The Penguin Press, 2009), p. 33.

[106] Congressional Budget Office, *Policy Options for Unmanned Aircraft Systems*, Pub. No. 4083, June 2011, p. 5.

[107] Glenn W. Goodman, Jr., "UAVs Come of Age," *The ISR Journal*, July 2002, p. 24.

[108] Department of Defense, *Selected Acquisition Report (SAR), MQ-1C UAS Gray (sic) Eagle*, DD-A&T(Q&A)823-420, Washington, DC, December 31, 2010.

missiles. Each platoon includes two ground control stations, two ground data terminals, one satellite communication ground data terminal, one portable ground control station, one portable ground data terminal, an automated takeoff and landing system, two tactical automatic landing systems, and ground support equipment. In total, the program will be 124 aircraft, plus 21 attrition aircraft and 7 schoolhouse aircraft, for a total of 152 aircraft. The average procurement unit cost of a Grey Eagle system is $114.1 million.[109]

MQ-9 Reaper

The MQ-9 Reaper, formerly the "Predator B," is General Atomics' follow-on to the MQ-1. The Reaper is a medium- to high-altitude, long-endurance Predator optimized for surveillance, target acquisition, and armed engagement. While the Reaper borrows from the overall design of the Predator, the Reaper is 13 feet longer and carries a 16-foot-longer wingspan. It also features a 900 hp turboprop engine, which is significantly more powerful than the Predator's 115 hp engine. These upgrades allow the Reaper to reach a maximum altitude of 50,000 feet, a maximum speed of 225 knots, a maximum endurance of 32 hours, and a maximum range of 2,000 nautical miles.[110] However, the feature that most differentiates Reaper from its predecessor is its ordnance capacity. While the Predator is outfitted to carry 2 100-pound Hellfire missiles, the Reaper now can carry as many as 16 Hellfires, equivalent to the Army's Apache helicopter, or a mix of 500-pound weapons and Small Diameter Bombs.

As of February 4, 2011, General Atomics Aeronautical Systems had delivered 65 of 399 planned Reapers, 43 of which are operationally active.[111]

The MQ-9 is operated by the 17th Reconnaissance Squadron and the 42nd Attack Squadron, both at Creech Air Force Base, NV, and the 29th Attack Squadron at Holloman AFB, NM.[112]

Program Status. Predator–family UAS are operated as part of a system, which consists of four air vehicles, a ground control station, and a primary satellite link. The unit cost in FY2009 for one Predator system was approximately $20 million,[113] while the average procurement unit cost for a Reaper system was $26.8 million.[114]

[109] Ibid.

[110] OSD, *UAS Roadmap 2005-2030*, August 2005, p. 10.

[111] Department of Defense, *Selected Acquisition Report (SAR), MQ-9 UAS Reaper*, DD-A&T(Q&A)823-424, Washington, DC, December 31, 2010.

[112] U.S. Air Force, *Fact Sheet: MQ-9 Reaper*, August 18, 2010.

[113] U.S. Air Force, *Fact Sheet: MQ-1 Predator*, July 20, 2010.

[114] Department of Defense, *Selected Acquisition Report (SAR), MQ-9 UAS Reaper*, DD-A&T(Q&A)823-424, Washington, DC, December 31, 2010.

Table 8. Predator and Reaper Combined Funding

($ in Millions)

		Procurement		RDT&E
FY11				
	Request	62 air vehicles	1600.0	
		Mods	31.8	
	Appropriations Conference	12 air vehicles	176.6	83.2
		Mods	31.8	
FY12				
	Request	9 air vehicles	125.5	61
		Mods	30.2	

RQ-4 Global Hawk

Northrop Grumman's RQ-4 Global Hawk has gained distinction as the largest and most expensive UAS currently in operation for the Department of Defense. Global Hawk incorporates a diverse surveillance payload with performance capabilities that rival or exceed most manned spy planes. However, Pentagon officials and Members of Congress have become increasingly concerned with the program's burgeoning cost, which resulted in Nunn-McCurdy breaches in April 2005 and April 2011.[115] Also, the RQ-4B Block 30 was deemed "not operationally suitable" due to "low air vehicle reliability" by the office of Operational Test and Evaluation in May 2011.[116]

System Characteristics. At 44 feet long and weighing 26,750 lbs, Global Hawk is about as large as a medium sized corporate jet. Global Hawk flies at nearly twice the altitude of commercial airliners and can stay aloft at 65,000 feet for as long as 35 hours. It can fly to a target area 5,400 nautical miles away, loiter at 60,000 feet while monitoring an area the size of the state of Illinois for 24 hours, and then return. Global Hawk was originally designed to be an autonomous drone capable of taking off, flying, and landing on pre-programmed inputs to the UAV's flight computer. Air Force operators have found, however, that the UAS requires frequent intervention by remote operators.[117] The RQ-4B resembles the RQ-4A, yet features a significantly larger airframe. In designing the B-model, Northrop Grumman increased the Global Hawk's length from 44 feet to 48 feet and its wingspan from 116 feet to 132 feet. The expanded size enables the RQ-4B to carry an extra 1000 pounds of surveillance payload.

[115] Amy Butler, "USAF Declares Second Major Global Hawk Cost Breach," Aerospace Daily, April 13, 2011. The 2005 breach stemmed from costs in transitioning from the Block 10 Global Hawk to the larger Block 30; the 2011 breach was attributed primarily to reduced procurement quantities rather than issues with the program.

[116] J. Michael Gilmore, *RQ-4B Global Hawk Block 30* , OSD Director, Operational Test and Evaluation, Operational Test and Evaluation Report, May 2011. John T. Bennett, "Pentagon testers slam aerial spy drone as unfit for operations," *The Hill*, June 7, 2011.

[117] Jeff Morrison, "USAF No Longer Viewing Global Hawk As An Autonomous System, Official Says," *Aerospace Daily*, December 3rd, 2005.

Mission and Payload. The Global Hawk UAS has been called "the theater commander's around-the-clock, low-hanging (surveillance) satellite."[118] The UAS provides a long-dwell presence over the battlespace, giving military commanders a persistent source of high-quality imagery that has proven valuable in surveillance and interdiction operations. The RQ-4A's current imagery payload consists of a 2,000-lb integrated suite of sensors much larger than those found on the Predator. These sensors include an all-weather SAR with Moving Target Indicator (MTI) capability, an E-O digital camera and an IR sensor. As the result of a January 2002 Air Force requirements summit, Northrop Grumman expanded its payload to make it a multi-intelligence air vehicle. The subsequent incarnation, the RQ-4B, is outfitted with an open-system architecture that enables the vehicle to carry multiple payloads, such as signals intelligence (SIGINT) and electronic intelligence (ELINT) sensors. Furthermore, the classified Multi-Platform Radar Technology Insertion Program (MP-RTIP) payload will be added in order to increase radar capabilities. These new sensor packages will enable operators to eavesdrop on radio transmissions or to identify enemy radar from extremely high altitudes. Future plans include adding hyper-spectral sensors for increased imagery precision and incorporating laser communications to expand information transfer capabilities.[119] The end goal is to field a UAS that will work with space-based sensors to create a "staring net" that will prevent enemies from establishing a tactical surprise.[120] In August 2003, the Federal Aviation Administration granted the Global Hawk authorization to fly in U.S. civilian airspace, which further expanded the system's mission potential.[121] This distinction, in combination with the diverse surveillance capabilities, has led many officials outside the Pentagon to consider the Global Hawk an attractive candidate for anti-drug smuggling and Coast Guard operations.[122]

Program Status. Developed by Northrop Grumman Corporation of Palmdale, CA, Global Hawk entered low-rate initial production in February 2002. The Air Force has stated that it intends to acquire 51 Global Hawks, at an expected cost of $6.6 billion for development and procurement costs. As of November 2009, the Air Force possessed 7 RQ-4As and 3 RQ-4Bs.[123] Another 32 Global Hawks had been authorized and appropriated through FY2011.[124] According to the most recent Selected Acquisition Report, the current average procurement unit cost for the Global Hawk has reached $140.9 million in current dollars.[125]

In April 2005, the Air Force reported to Congress that the program had overrun by 18% as a result of an "increasing aircraft capacity to accommodate requirements for a more sophisticated, integrated imagery and signals intelligence senor suite."[126] A Government Accountability Office report in December 2004 noted that the program had increased by nearly $900 million since 2001 and recommended delaying the purchase of future Global Hawks until an appropriate

[118] Glenn W. Goodman, Jr., "UAVs Come Of Age," *The ISR Journal*, July 2002.

[119] David A. Fulghum, "Global Hawk Shows Off Updated Package of Sensors," *Aviation Week & Space Technology*, September 08, 2003.

[120] Ibid.

[121] Sue Baker, "FAA Authorizes Global Hawk Flights," Aeronautical Systems Center Public Affairs, August 21, 2003.

[122] Ron Laurenzo, "Global Hawk Scouts Ahead for Other UAVs," *Defense Week*, September 2, 2003.

[123] U.S. Air Force, *Fact Sheet: RQ-4 Global Hawk*, November 19, 2009.

[124] Department of Defense, *Department of Defense Fiscal Year 2012 Budget Estimates, Aircraft Procurement, Air Force*, February 2011.

[125] OSD, *Selected Acquisition Report*, December 31, 2010, p. 29.

[126] James R. Asker, "Global Hawk 18% Over Budget," *Aviation Week & Space Technology*, April 25, 2005.

development strategy could be implemented.[127] The rising costs of the UAV and accusations of Air Force mismanagement have caused concern among many in Congress and in the Pentagon as well as facilitating an overall debate on the Air Force's development strategy.[128]

Following a 2010 Defense Acquisition Board review of the Global Hawk program,

> Air Force acquisition executive David Van Buren told reporters that he is "not happy" with the pace of the program, both on the government and the contractor side. Chief Pentagon arms buyer Ashton Carter also criticized the program, saying that it was "on a path to being unaffordable."[129]

In April 2011, a reduction in the number of Global Hawk Block 40 aircraft requested in the FY2012 budget from 22 to 11 caused overall Global Hawk unit prices to increase by 11%, again triggering Nunn-McCurdy.[130]

In its markup of the FY2011 defense authorization bill, the House Armed Services Committee expressed concern "that differing, evolving service unique requirements, coupled with Global Hawk UAS vanishing vendor issues, are resulting in a divergence in each service's basic goal of maximum system commonality and interoperability, particularly with regard to the communications systems." The bill report directs the Under Secretary of Defense for Acquisition, Technology, and Logistics to certify and provide written notification to the congressional defense committees by March 31, 2011, that he has reviewed the communications requirements and acquisition strategies for both Global Hawk and BAMS. The subcommittee wants assurance that the requirements for each service's communications systems have been validated and that the acquisition strategy for each system "achieves the greatest possible commonality and represents the most cost effective option" for each program.[131]

A May 20, 2011, report from the Air Force Operational Test and Evaluation Center found the Global Hawk Block 20/30 to be "effective with significant limitations ... not suitable and partially mission capable." The report cited "lackluster performance of the EISS imagery collector and ASIP sigint collectors at range" rather than issues with the Global Hawk airframe itself.[132]

---

[127] United States Government Accountability Office, GAO-05-6 *Unmanned Aerial Vehicles[:] Changes in Global Hawk's Acquisition Strategy Are Needed to Reduce Program Risks,* November 2004, p. 3-4.

[128] See H.Rept. 109-89. House Armed Service Committee "National Defense Authorization Act for the Fiscal Year 2006." May 20, 2005, p. 91.

[129] Marina Malenic, "Air Force, Navy Pledge Greater Global Hawk-BAMS Cooperation," *Defense Daily*, July 2, 2010.

[130] The Nunn-McCurdy provision requires DOD to notify Congress when cost growth on a major acquisition program reaches 15%. If the cost growth hits 25%, Nunn-McCurdy requires DOD to justify continuing the program based on three main criteria: its importance to U.S. national security; the lack of a viable alternative; and evidence that the problems that led to the cost growth are under control. For more information, see CRS Report R41293, *The Nunn-McCurdy Act: Background, Analysis, and Issues for Congress* , by Moshe Schwartz.

[131] U.S. Congress, House Committee on Armed Services, *National Defense Authorization Act for Fiscal Year 2011*, Report to accompany H.R. 5136, 111th Cong., May 21, 2010, H.Rept. 111-491, p. 178.

[132] Amy Butler, "Poor Testing Results Latest Hurdle for Global Hawk," *Aviation Week/Ares blog*, June 3, 2011.

Table 9. Global Hawk Funding

($ in Millions)

	Quantity	Procurement	RDT&E	Advance Procurement
FY11				
Request	4	649.6	251.3	90.2
Authorization Conference[a]	N/A	N/A	N/A	N/A
Appropriations Conference	4	503.0	220.3	72.3
FY12				
Request	3	323.9		71.5

a. As passed, H.R. 6523, the Ike Skelton National Defense Authorization Act For Fiscal Year 2011, did not include program-level detail, so no amounts were specified for these program elements.

BAMS

The Navy's Broad Area Maritime Surveillance system is based on the Global Hawk Block 20 airframe but with significantly different sensors from its Air Force kin. This, coupled with a smaller fleet size, results in a higher unit cost. "The air service's drone costs $27.6 million per copy, compared to an expected $55 million per BAMS UAV, including its sensors and communications suite.... At 68 aircraft, the BAMS fleet will be the world's largest purchase of long-endurance marinized UAVs."[133]

System Characteristics and Mission. "BAMS ... provides persistent maritime intelligence, surveillance, and reconnaissance data collection and dissemination capability to the Maritime Patrol and Reconnaissance Force. The MQ-4C BAMS UAS is a multi-mission system to support strike, signals intelligence, and communications relay as an adjunct to the MMA/P-3 community to enhance manpower, training and maintenance efficiencies worldwide."[134]

"The RQ-4 ... features sensors designed to provide near worldwide coverage through a network of five orbits inside and outside continental United States, with sufficient air vehicles to remain airborne for 24 hours a day, 7 days a week, out to ranges of 2000 nautical miles. Onboard sensors will provide detection, classification, tracking and identification of maritime targets and include maritime radar, electro-optical/infra-red and Electronic Support Measures systems. Additionally, the RQ-4 will have a communications relay capability designed to link dispersed forces in the theater of operations and serve as a node in the Navy's FORCEnet strategy."[135]

[133] Gayle S. Putrich, "Northrop selected to build BAMS drone," *Navy Times*, April 22, 2008.

[134] Northrop Grumman, "MQ-4C BAMS UAS," press release, http://www.as.northropgrumman.com/products/bams/index.html.

[135] Department of Defense, *Department of Defense Fiscal Year 2012 Budget Estimates, Research, Development, Test & Evaluation, Navy*, Budget Activity 7, February 2011.

"The drones ... will collect information on enemies, do battle-damage assessments, conduct port surveillance and provide support to Navy forces at sea. Each aircraft is expected to serve for 20 years."[136]

Program Status. The Administration's FY2012 budget request documents place Milestone C for BAMS in the third quarter of FY2013, with initial operational capability in the first quarter of 2016. "Since Milestone B for the Navy BAMS UAS program, identifying opportunities for the RQ-4-based BAMS and Global-Hawk programs has been a significant interest item for the UAS TF and has been well documented within the Department."[137] In one effort to integrate development, on June 12, 2010, the Navy and Air Force concluded a Memorandum of Agreement (MOA) regarding their Global Hawk and BAMS programs, which use a common airframe. "Shared basing, maintenance, command and control, training, logistics and data exploitation are areas that could be ripe for efficiencies, says Lt. Gen David Deptula, Air Force deputy chief of staff for intelligence, surveillance and reconnaissance.... Also, a single pilot and maintenance training program is being established at Beale AFB, Calif., for both fleets."[138] However, issues still exist over common control stations and whether one service's pilots should be able to operate the other service's aircraft.

MQ-8B Fire Scout

Now in deployment, the Fire Scout was initially designed as the Navy's choice for an unmanned helicopter capable of reconnaissance, situational awareness, and precise targeting.[139] Although the Navy canceled production of the Fire Scout in 2001, Northrop Grumman's vertical take-off UAV was rejuvenated by the Army in 2003, when the Army designated the Fire Scout as the interim Class IV UAV for the future combat system. The Army's interest spurred renewed Navy funding for the MQ-8, making the Fire Scout DOD's first joint UAS helicopter.

System Characteristics and Mission. Northrop Grumman based the design of the Fire Scout on a commercial helicopter. The RQ-8B model added a four-blade rotor to reduce the aircraft's acoustic signature.[140] With a basic 127-pound payload, the Fire Scout can stay aloft for up to 9.5 hours; with the full-capacity sensor payload, endurance diminishes to roughly 6 hours. Fire Scout possesses autonomous flight capabilities. The surveillance payload consists of a laser designator and range finder, an IR camera and a multi-color EO camera, which when adjusted with specific filters could provide mine-detection capabilities.[141] Fire Scout also currently possesses line-of-sight communication data links. Initial tests of an armed Fire Scout were conducted in 2005, and the Navy expects to add "either Raytheon's Griffin or BAE's Advanced Precision Kill Weapon System" small missiles to currently deployed Fire Scouts soon.[142] Discussions of future missions

---

[136] Gayle S. Putrich, "Northrop selected to build BAMS drone," *Navy Times*, April 22, 2008.

[137] Under Secretary of Defense (Acquisition, Technology and Logistics), *Department of Defense Report to Congress on Addressing Challenges for Unmanned Aircraft Systems*, September 2010.

[138] Amy Butler, "U.S. Navy/Air Force UAV Agreement Raises Questions," *Aerospace Daily*, July 6, 2010.

[139] "RQ-8A Fire Scout, Vertical Take Off and Landing Tactical Unmanned Aerial Vehicle (VTUAV)," GlobalSecurity.org, http://www.globalsecurity.org/intell/systems/vtuav.htm, April 26th, 2004.

[140] David A. Fulghum, "Army Adopts Northrop Grumman's Helicopter UAV," *Aviation Week & Space Technology*, October 20th, 2003.

[141] Northrop Grumman Corp. Press Release, "Northrop Grumman's Next-Generation Fire Scout UAV on Track," June 23rd, 2005.

[142] Joshua Stewart, "Navy plans to arm Fire Scout UAV with missiles," *Navy Times*, August 18, 2011.

have also covered border patrol, search and rescue operations, medical resupply, and submarine spotting operations.

Program Status. Six production MQ-8 air vehicles have been delivered to date.[143] The Pentagon's 2009 UAS Roadmap estimates a future inventory of 131 RQ-8Bs for the Navy to support the Littoral Combat Ship class of surface vessels.[144] The Army had intended to use the Fire Scout as the interim brigade-level UAV for its Future Combat System program,[145] but canceled its participation in January, 2010.[146]

A Fire Scout attracted media attention in August 2010, when it flew through Washington, DC, airspace after losing its control link. "A half-hour later, Navy spokesmen said, operators re-established control and the drone landed safely."[147]

FIRE-X/MQ-8C

The FIRE-X project, recently designated MQ-8C but continuing the Fire Scout name, is a developmental effort to adapt the Fire Scout software and navigation systems to a full-size standard helicopter. The Navy "is to award Northrop Grumman a contract to supply 28 MQ-8C Fire Scout ... to be fielded by the first quarter of 2014 to meet an urgent operational requirement."[148]

> Fire Scout can fly for 8 hours with a maximum range of 618 nautical miles? Well, Fire-X will fly for 15, with a max range of 1227. Fire Scout tops out at 100 knots? Fire-X can speed by at 140. Fire-X will carry a load of 3200 lbs. to Fire Scout's 1242. All this talk from a drone helicopter that just took its first flight in December.... Fire-X isn't going to be a big departure from Fire Scout, though. The BRITE STAR II and other radars will remain on board, as will its software for relaying information to a ship.[149]

RQ-170 Sentinel

Although publicly acknowledged to exist, most information about the Lockheed Martin RQ-170 Sentinel is classified. First photographed in the skies over Afghanistan, but also reportedly in operation from South Korea,[150] the RQ-170 is a tailless "flying wing" stealthier than other current

[143] Department of the Navy, *Fiscal Year (FY) 2012 Budget Estimates, MQ-8 UAV*, February 2011.

[144] Northrop Grumman Corp. Press Release, "Northrop Grumman's Next-Generation Fire Scout UAV on Track," June 23rd, 2005. *FY2009–2034 Unmanned Systems Integrated Roadmap*, p. 66.

[145] The Army intended to field four different classes of UAVs as part of its Future Combat System (FCS): Class I for platoons, Class II for companies, Class III for battalions, and Class IV for brigades. See CRS Report RL32888, *Army Future Combat System (FCS) "Spin-Outs" and Ground Combat Vehicle (GCV): Background and Issues for Congress*, by Andrew Feickert and Nathan J. Lucas, for more information.

[146] April M. Havens, "U.S. Army wants to cancel Fire Scouts (sic) program," *The Mississippi Press*, January 15, 2010, As presented on http://www.gulflive.com.

[147] Elisabeth Bumiller, "Navy Drone Violated Washington Airspace," *The New York Times*, August 26, 2010.

[148] Graham Warwick, "U.S. Navy Goes Ahead With Bell 407-Based Fire Scout UAV," *Aerospace Daily*, September 8, 2011.

[149] Spencer Ackerman, "Navy Upgrades Its Spying, Drug-Sniffing Robot Copter," *Wired.com/Danger Room blog*, April 11, 2011.

[150] Bill Sweetman, "Beast Sighted In Korea," *Aviation Week/Ares blog*, February 16, 2010.

U.S. UAS. An RQ-170 was reported to have performed surveillance and data relay related to the operation against Osama bin Laden's compound on May 1, 2011. The government of Iran claimed on December 2, 2011, to be in possession of an intact RQ-170 following its incursion into Iranian airspace.

System Characteristics. Built by Lockheed Martin, the RQ-170 has a wingspan of about 65 feet and is powered by a single jet engine. It appears to have two sensor bays (or satellite dish enclosures) on the upper wing surface. Although an inherently low-observable blended wing/fuselage design like the B-2, the RQ-170's conventional inlet, exhaust, and landing gear doors suggest a design not fully optimized for stealth.[151]

Potential Mission and Payload. "The RQ-170 will directly support combatant commander needs for intelligence, surveillance and reconnaissance to locate targets."[152]

Program Status. "The RQ-170 is a low observable unmanned aircraft system (UAS) being developed, tested and fielded by the Air Force."[153] No further official status is available.

Other Current UAS Programs

RQ-5A Hunter/MQ-5B Hunter II

Originally co-developed by Israel Aircraft Industries and TRW (now owned by Northrop Grumman) for a joint U.S. Army/Navy/Marine Corps short-range UAS, the Hunter system found a home as one of the Army's principal unmanned platforms. The service has deployed the RQ-5A for tactical ISR in support of numerous ground operations around the world. At one time, the Army planned to acquire 52 Hunter integrated systems of eight air vehicles apiece, but the Hunter program experienced some turbulence. The Army canceled full-rate production of the RQ-5A in 1996, but continued to use the seven systems already produced. It acquired 18 MQ-5B Hunter IIs through low-rate initial production in FY2004 and FY2005. The MQ-5B's design includes longer endurance and the capability to be outfitted with anti-tank munitions. Both variants are currently operated by the 224th Military Intelligence Battalion out of Fort Stewart, GA; by the 15th Military Intelligence Battalion out of Ft. Hood, TX; and by 1st Military Intelligence Battalion out of Hohenfels, Germany.

System Characteristics. The RQ-5A can fly at altitudes up to 15,000 feet, reach speeds of 106 knots, and spend up to 12 hours in the air. Weighing 1,600 pounds, it has an operating radius of 144 nautical miles. The MQ-5B includes an elongated wingspan of 34.3 feet up from 29.2 feet of the RQ-5A and a more powerful engine, which allows the Hunter II to stay airborne for three extra hours and to reach altitudes of 18,000 feet.[154] The Hunter system consists of eight aircraft, ground control systems and support devices, and launch/recovery equipment. In FY2004, the final year of Hunter procurement, a Hunter system cost $26.5 million.

[151] U.S. Air Force, *Fact Sheet: RQ-170 Sentinel*, December 2, 2010. and CRS analysis of available RQ-170 photography.

[152] U.S. Air Force, *Fact Sheet: RQ-170 Sentinel*, December 2, 2010.

[153] Ibid.

[154] OSD, *UAS Roadmap 2005-2030*, August 2005, p. 7

Mission and Payload. The Army has mostly used the Hunter system for short- and medium-range surveillance and reconnaissance. More recently, however, the Army expanded the Hunter's missions, including weaponization for tactical reconnaissance/strike operations with the GBU-44/B Viper Strike precision guided munition, which can designate targets either from the munition's laser, from another aerial platform, or from a ground system. This weapon makes the Hunter the Army's first armed UAS. "Also, in 2004, the Department of Homeland Security, Customs and Border Protection Bureau, and Office of Air and Marine utilized Hunter under a trial program for border patrol duties. During this program, the Hunter flew 329 flight hours, resulting in 556 detections."[155]

Program Status. The Army halted Hunter production in 2005. As of May 2011, 45 Hunter UAVs were still in operation and periodically receiving upgrades and modifications. In August 2005, the Army awarded General Atomics' Warrior UAS (which later became Grey Eagle) the contract for the Extended Range-Multi Purpose UAS program over the Hunter II.[156]

RQ-7 Shadow

The RQ-7 Shadow found a home when the Army, after a two-decade search for a suitable system, selected AAI's close range surveillance platform for its tactical unmanned aerial vehicle (TUAV) program. Originally, the Army, in conjunction with the Navy explored several different UAVs for the TUAV program, including the now-cancelled RQ-6 Outrider system. However, in 1997, after the Navy pursued other alternatives, the Army opted for the low-cost, simple design of the RQ-7 Shadow 200. Having reached full production capacity and an IOC in 2002, the Shadow has become the primary airborne ISR tool of numerous Army units around the world and is expected to remain in service through the decade.

The Administration's FY2011 budget request did not include funding for Shadow aircraft, although it did include continued RDT&E funding for Shadow.[157]

System Characteristics. Built by AAI Corporation (now owned by Textron), the Shadow is 11 feet long with a wingspan of 13 feet. It has a range of 68 nautical miles, a distance picked to match typical Army brigade operations, and average flight duration of five hours. Although the Shadow can reach a maximum altitude of 14,000 feet, its optimum level is 8,000 feet. The Shadow is catapulted from a rail-launcher, and recovered with the aid of arresting gear. The UAS also possesses automatic takeoff and landing capabilities. The upgraded version, the RQ-7B Shadow, features a 16-inch greater wingspan and larger fuel capacity, allowing for an extra two hours of flight endurance.

Mission and Payload. The Shadow provides real-time reconnaissance, surveillance, and target acquisition information to the Army at the brigade level. A potential mission for the Shadow is the perilous job of medical resupply. The Army is considering expanding the UAS's traditional missions to include a medical role, where several crucial items such as blood, vaccines, and fluid

[155] Kari Hawkins, "Pioneer platform soars to battlefield success," *www.army.mil*, May 19, 2011.

[156] Jefferson Morris, "Army More Than Doubles Expected Order for ERMP with General Atomics Win," *Aerospace Daily & Defense Report*, August 10th, 2005.

[157] Todd Harrison, *Analysis of the FY 2011 Defense Budget* (Washington, DC: Center for Strategic & Budgetary Assessments, 2010), p. 38.

infusion systems could be delivered to troops via parachute.[158] For surveillance purposes, the Shadow's 60-pound payload consists of an E-O/IR sensor turret, which produces day or night video and can relay data to a ground station in real-time via a line-of-sight data link. As part of the Army's Future Combat System plans, the Shadow will be outfitted with the Tactical Common Data Link currently in development to network the UAS with battalion commanders, ground units, and other air vehicles.[159] The Marine Corps is considering how to arm Shadow.[160]

Program Status. The Army and Marine Corps currently maintain an inventory of 364 Shadow UAVs.[161] The program cost for a Shadow UAV system—which includes four vehicles, ground control equipment, launch and recovery devices, remote video terminals, and High Mobility Multipurpose Wheeled Vehicles for transportation—reached $11.1 million in current year dollars for FY2008.[162] The Army procured 102 systems through 2009.[163] In FY2012, the Army requested $25 million for 20 Shadow aircraft to replace combat losses, and approximately $200 million for payload upgrades.

"Small UAVs"

RQ-14 Dragon Eye

AeroVironment's Dragon Eye is a backpack-carried, battery-operated UAV employed by the Marines at the company level and below for reconnaissance, surveillance, and target acquisition. Dragon Eye features a 3.8-foot rectangular wing, twin propellers, and two camera ports each capable of supporting day-light electro-optical cameras, low-light TV cameras, and infrared cameras. The compact and lightweight design of the UAV allows an operational endurance of 45 minutes and can travel as far as 2.5 nautical miles from the operator. Low-rate-initial-production of 40 aircraft began in 2001. After a 2003 operational assessment, the Marine Corps awarded AeroVironment a contract to deliver approximately 300 systems of full-rate-production Dragon Eyes.[164] However, that contract was later revised to acquire Raven UAS instead. One Dragon Eye system consists of three air vehicles and one ground station. The final Marine Corps procurement budget request in FY2006 anticipated the current unit cost per Dragon Eye system as $154,000.[165]

FQM-151 Pointer

Although procurement of this early UAS began in 1990, the electric-powered Pointer has seen service in Operation Enduring Freedom (OEF) and Operation Iraqi Freedom (OIF). Pointer is a

[158] Erin Q. Winograd, "Army Eyes Shadow UAVs Potential For Medical Resupply Missions," *InsideDefense.com*, December 20, 2002, p.14.

[159] "Upgraded Shadow UAV Rolls Off Production Line," *Defense Today*, August 5, 2004.

[160] Paul McCleary, "Marines Want a Big Bang From a Small Package," *Aviation Week/Ares blog*, October 26, 2010.

[161] OSD, *UAS Roadmap 2005-2030*, August 2005, p. 8.

[162] Department of the Army, Army Procurement OPA 02: Communications and Electronics FY2009, February 2008, TUAS (B00301), Item No. 61, p. 18 of 23.

[163] Department of the Army, *Committee Staff Procurement Backup Book, Fiscal Year 2012 Budget Estimate, Aircraft Procurement, Army, RQ-7 UAV MODS*, Washington, DC, February 2011, p. 1 of 10.

[164] Peter La Franchi, "Directory: Unmanned Air Vehicles," *Flight International*, June 21st, 2005, p. 56.

[165] Department of the Navy, FY2006-FY2007 Budget Estimate - Marine Corps Procurement, February 2005, BLI No. 474700, Item 44, p. 20 of 22.

short-range reconnaissance and battlefield surveillance UAV developed by AeroVironment. Its flight endurance (two hours) is greater than most similar small UAVs, in part due to its relatively large, 9-foot wingspan. That wingspan decreases portability of the 8.5-pound Pointer, and as a result, transportation of a Pointer system (two air vehicles and a ground control unit) requires two personnel. Although superseded by the Raven (below), Pointer remains a valued short-range ISR asset for the Air Force and Special Operations Command.

RQ-11 Raven

Engineered from the basic design of the Pointer, the Raven is two-thirds the size and weight of its predecessor, with a much smaller control station, making the system man-portable.[166] "The RQ-11A is essentially a down-sized FQM-151 *Pointer*, but thanks to improved technology can carry the same navigation system, control equipment, and payload."[167] The Raven provides Army and SOCOM personnel with "over-the-hill" reconnaissance, sniper spotting, and surveillance scouting of intended convoy routes. The electric motor initiates flight once hand-launched by a running start from the ground operator. The vehicle is powered by an electric battery that needs to be recharged after 90 minutes, but deployed soldiers are equipped with four auxiliary batteries that can be easily charged using the 28 volt DC outlet in a Humvee. The vehicle lands via a controlled crash in which the camera separates from the body, which is composed of Kevlar plating for extra protection. Like the Pointer, the Raven can carry either an IR or an E-O camera and transmits real-time images to its ground operators. The relatively simple system allows soldiers to be trained in-theater in a matter of days. Raven systems can either be deployed in three-aircraft or two-aircraft configurations. "Raven was adopted as the US Army's standardised short range UAV system in 2004 with a total of 2469 air vehicles (including older RQ-11A series models) in operational service by mid 2007."[168] "The US Army has an ongoing acquisition objective for about 2,200 Raven systems and has taken delivery of more than 1,300 to date."[169] A three-aircraft system costs approximately $167,000.[170]

ScanEagle

Developed by the Insitu Group (owned by Boeing) as a "launch-and-forget" UAV, the ScanEagle autonomously flies to points of interest selected by a ground operator.[171] The ScanEagle has gained notice for its long endurance capabilities and relative low cost. The gasoline-powered UAV features narrow 10 foot wings that allow the 40-pound vehicle to reach altitudes as high as 19,000 feet, distances of more than 60 nautical miles, and a flight endurance of almost 20 hours. Using an inertially stabilized camera turret carrying both electro-optical and infrared sensors, ScanEagle currently provides Marine Corps units in Iraq with force-protection ISR and is also used by Special Operations Command. ScanEagle operations began in 2004,[172] and continue

---

[166] Peter La Franchi, "Directory: Unmanned Air Vehicles," *Flight International*, June 21st, 2005, p. 57.

[167] Andreas Parsch, "AeroVironment RQ-11 Raven," *Directory of U.S. Military Rockets and Missiles*, September 12, 2006.

[168] "UAV Directory - Aircraft Specification AeroVironment - RQ-11A Raven," *FlightGlobal.com*, (2011).

[169] "UAV Directory - Aircraft Specification AeroVironment - RQ-11B Raven," *FlightGlobal.com*, (2011).

[170] The FY2012 budget request includes $70.8 million for 424 systems and supporting equipment. Department of Defense, *Department of Defense Fiscal Year 2012 Budget Estimates, Aircraft Procurement, Army*, February 2011.

[171] Jim Garamone, "ScanEagle Proves Worth in Fallujah Fight," *American Forces Press Service*, January 11th, 2005.

[172] Insitu, "Backgrounder: ScanEagle® Unmanned Aircraft System," press release, September 22, 2011, (continued...)

today. Although ScanEagle was expected to cost about $100,000 per copy, the Navy and SOCOM have contracted for operations instead of procurement, with Boeing providing ISR services utilizing ScanEagle under a fee-for-service arrangement.[173]

ScanEagle is also in use by non-military organizations for surveillance purposes, including tracking whale migrations.[174]

Small Tactical Unmanned Aerial System (STUAS)

In July 2010, the Department of the Navy awarded Insitu a two-year, $43.7 million contract for the design, development, integration, and test of the Small Tactical Unmanned Aircraft System (STUAS) for use by the Navy and Marine Corps to provide persistent maritime and land-based tactical reconnaissance, surveillance, and target acquisition (RSTA) data collection and dissemination. "For the USMC, STUAS will provide the Marine Expeditionary Force and subordinate commands (divisions and regiments) a dedicated ISR system capable of delivering intelligence products directly to the tactical commander in real time. For the Navy, STUAS will provide persistent RSTA support for tactical maneuver decisions and unit-level force defense/force protection for Navy ships, Marine Corps land forces, and Navy Special Warfare Units."[175]

Payloads include day/night video cameras, an infrared marker, and a laser range finder, among others. STUAS can be launched and recovered from an unimproved expeditionary/urban environment, as well as from the deck of Navy ships.[176]

STUAS uses Insitu's Integrator airframe, which uses common launch, control, and recovery equipment with ScanEagle. STUAS has a takeoff weight of up to 125 pounds with a range of 50 nautical miles. However, STUAS will be procured and operated by the services rather than operated on a fee-for-service basis because "the Scan Eagle's current fee-for-service contract limits the way the UAS is deployed ... with Boeing/Insitu employees usually operating the aircraft in the field due to liability issues." Procuring the system will allow the services to train their own operators. Initial operating capability is expected in the fourth quarter of FY2013.[177]

(...continued)

http://www.boeing.com/bds/mediakit/2011/ausa/pdf/bkgd_scaneagle.pdf.

[173] See, inter alia, Boeing, "Boeing Awarded Navy Contract for ScanEagle Services," press release, June 6, 2008, http://www.boeing.com/news/releases/2008/q2/080606a_nr.html and Insitu, "Boeing Wins $250M Special Ops Contract for ScanEagle ISR Services," press release, May 22, 2009, http://www.insitu.com/print.cfm?cid=3774.

[174] Boeing, " From saving soldiers to saving whales," undated press release, http://www.boeing.com.au/ViewContent.do?id=61784&aContent=ScanEagle.

[175] Naval Air Systems Command, *Aircraft and Weapons: Small Tactical Unmanned Aircraft System*, http://www.navair.navy.mil/index.cfm?fuseaction=home.display&key=4043B5FA-7056-4A3A-B038-C60B21641288.

[176] Ibid.

[177] Gayle Putrich, "Insitu wins long-awaited US Navy STUAS deal," *FlightGlobal.com*, July 30, 2010.

APPENDIX B

The Department of Defense's Traditional Acquisition System

Overview

The DAS is a structured and deliberate process for establishing requirements and planning and resourcing the procurement of new capabilities. The need for accountability and oversight in the allocation of national resources has led to the buildup over decades of a large bureaucratic apparatus around the DAS. The complexity of the process can be seen in Figure B.1.

This bureaucracy has many virtues. It provides for a structured process governed by regulations and accountable authorities and includes provisions for relatively stable funding over system life cycles. Such provisions are far from perfect. For example, observers have long noted that the process can be slow and unresponsive to the immediate needs of warfighters. Such concerns have led, as will be discussed below, to evolving roles for the combatant commanders in resource allocation processes. Another concern is that provision of training and sustainment considerations, while formally required, is an afterthought to the acquisition and fielding of new equipment. Both concerns will be discussed below.

Role of Combatant Commanders in Traditional Acquisition

To meet current warfighter needs, the DoD has in recent years advanced changes to some existing acquisition processes and relationships. As a result, the roles of the COCOMs in the traditional system have evolved. For example, the Secretary of Defense approved a "streamlined and refocused" integrated priority list process to better support the development of service program objective memorandums (Hicks et al., 2008, p. 59). In November 2005, the Vice Chairman of the Joint Chiefs of Staff launched a process by which the Joint Requirements Oversight Council, in consultation with the COCOMs, identified a list of "most pressing" military issues to support resourcing and planning decisions (Hicks et al., 2008, p. 59). In 2006 the Senior Warfighters Forum, consisting of COCOM deputies, began meeting every four to six weeks to discuss coordination issues, and the Vice Chairman of the Joint Chiefs of Staff began coordinating regular trips to the COCOMs with critical PPBE decision points to better represent the COCOM position in resourcing decisions (Hicks et al., 2008, p. 60). Furthermore, SOCOM's traditional acquisition activities have increased, and its budget has grown significantly (GAO, 2007b, p. 1).

The COCOMS do not lead the process, but their ability to participate in acquisition, particularly in identifying requirements, has expanded. An emphasis on warfighting in recent years has given new weight to the capabilities and solutions that the COCOMs identify as

Figure B.1
Defense Acquisition System

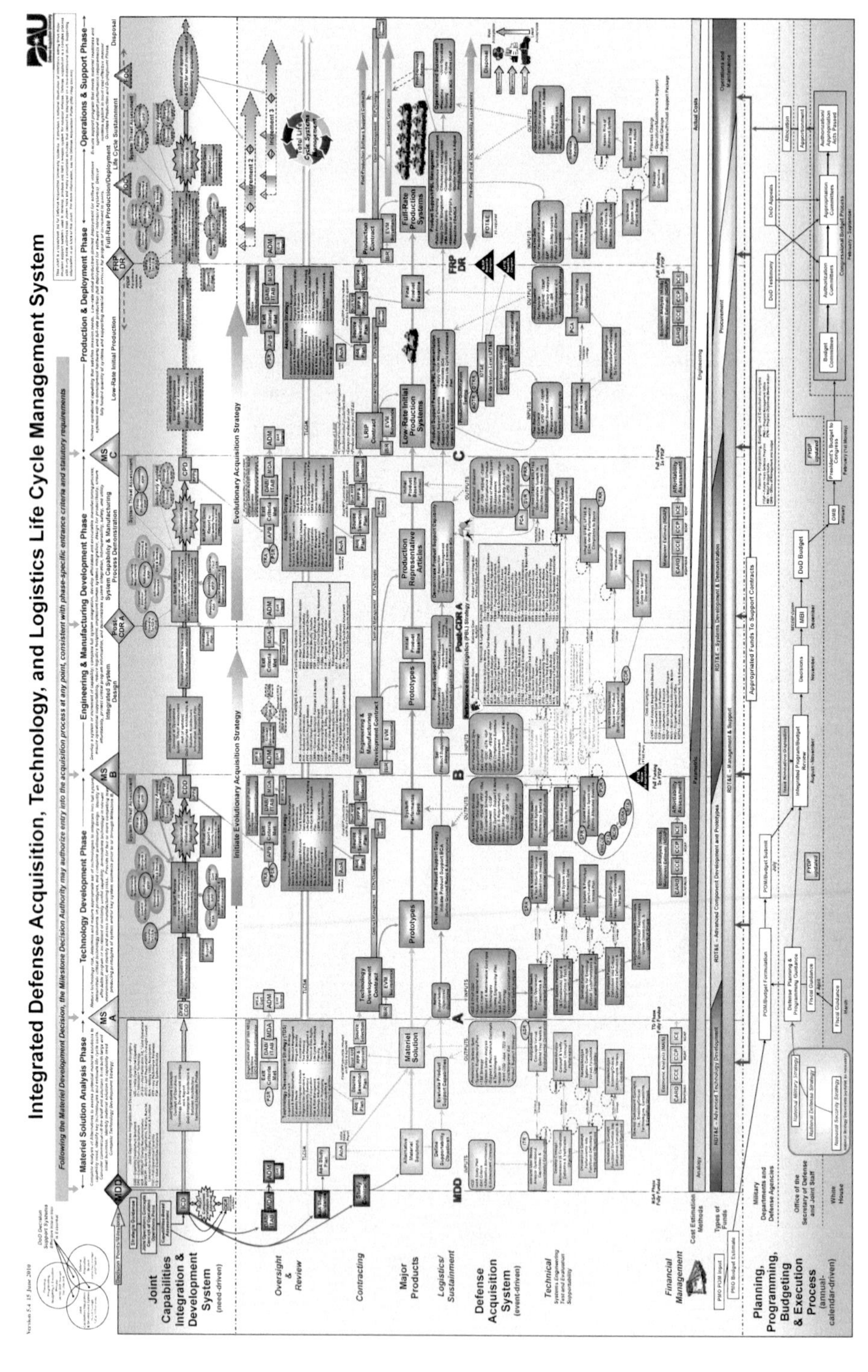

SOURCE: Defense Acquisition University, June 2010.

RAND *RR440-B.1*

required. Beyond the current fight, however, observers argue that the nature of future challenges and the technology necessary to meet them are so dynamic that fundamental acquisition relationships need to be rethought. As the commander of SOUTHCOM recently wrote:

> We are living in an age of rapid change facilitated by advancing technologies and increasingly networked systems, societies, and economies. In order for security agencies to be successful in this complex environment, those organizations must be flexible, open and forward-thinking. (Stavridis, 2010, p. xxii)

Leadership of traditional acquisition processes is vested in the military departments (the force providers) rather than the combatant commanders (the force users). The DAS consists of the interrelationship between the establishment of requirements (Joint Capabilities Integration and Development), the process of providing solutions, and the allocation of resources (PPBE). All processes are primarily driven by the military departments, with some inputs from the COCOMs. The logic behind the vesting of acquisition authority in the military departments reflects both Title 10 responsibilities and unique competencies for acquisition. The division of labor has traditionally been that the COCOMs provide inputs to the determination of requirements, and the services lead on providing solutions. A recent RAND assessment concluded that it "is the job of those in Washington, D.C., to reach out to the COCOMs and demonstrate that their needs are being addressed, rather than turn the process over to them" (Blickstein and Nemfakos, 2009, p. 7). Traditionally, the COCOMs are viewed as best suited to fighting wars, while the military departments are best suited to balancing other priorities and managing resource decisionmaking.

APPENDIX C

Military Value Analysis of Unmanned Aircraft System Training Bases

Training Locations

The information on specific training locations included in this appendix is drawn from RAND interviews, service presentations to congressional oversight committees, and Army surveys and consolidated snapshots of selected Army bases compiled in November 2012.

Initial training for UAS operators and sensor operators is concentrated at a few locations. The Air Force uses Beale AFB for training its Global Hawk crews and primarily Holloman AFB for its Predator crews.[1] The Navy plans to leverage the similarities between the Global Hawk and BAMS by having some joint training at Beale AFB. The Army conducts all its initial UAS training at Fort Huachuca, Arizona. The Marine Corps trains its Shadow crews with the Army.[2] Current and continuing pressures on the defense budget are likely to result in further changes. We limited our efforts to analyze military value in depth to activities that any training base would more certainly include, regardless of budget reductions. After we gain a fuller understanding of the results of Secretary of Defense decisions on the content of the FY 2014 and FY 2015 programs, we will renew our efforts to complete this military value analysis.

The approach to the *military value analysis of training bases* presented here is based on the Department of the Navy's 1995 Base Realignment and Closure documents. These categories and a brief description are listed below:

- flight training areas and airspace—covers access to special use airspace, availability of training areas, etc.
- airfield and maintenance facilities—includes facilities available for housing and maintaining the aircraft
- expansion potential—any comments on future plans for expansion or capacity for expansion
- training and training facilities—includes classrooms or other similar infrastructure and equipment required for training
- military and general support mission—any impacts of the other military missions housed at the same installation

[1] The training for the LRE is separate from the FTU and takes place at Creech AFB, Nevada.

[2] The Marines will also operate the RQ-20A Puma. We do not discuss the training for that SUAS.

- weather—any impacts of weather on training capacity
- location—description of the where the installation is located
- base loading—comments on consolidation of training.

Army RQ-7 and MQ-1C Training IQT Fort Huachuca, Arizona

Fort Huachuca, Arizona, is home to the 2-13th Aviation Regiment (formally the Unmanned Aircraft Systems Training Battalion). The 2-13th trains operators and maintainers for the RQ-7 Shadow, the Warrior Alpha, and the MQ-1C Gray Eagle. The training pipeline for each of these was described in the body of this report.

Flight Training Areas and Airspace

See Figure C.1. R-2303 is a major operational area used for airspace manned and unmanned operations, Army and joint service training along with Department of Homeland Security border security operations. Fort Huachuca is the controlling authority and so can activate this restricted use airspace on demand, at any time. It covers 850 mi^2 of land outside Fort Huachuca and encompasses 4,600 mi^3 of airspace from the surface to 30,000 feet, much of it over private land. The FAA has issued several waivers to ensure the safe operations of manned and unmanned traffic. The area also has unencumbered radio frequency spectrum for the foreseeable future.

Figure C.1
Airspace Surrounding Fort Huachuca

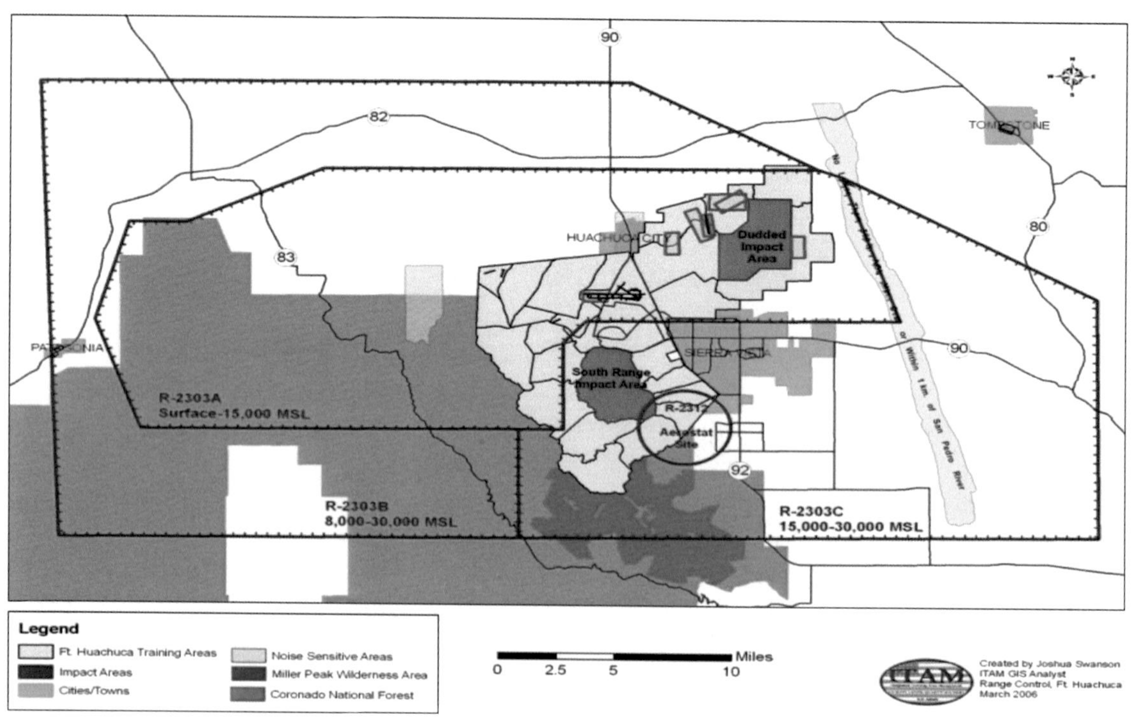

RAND *RR440-C.1*

Airfield and Maintenance Facilities

See Figure C.2. UAS operations split between the Libby Army Airfield and Black Tower. Shadow and Hunter training takes place at Black Tower. There are dedicated maintenance facilities as well as hangers and airstrips. Gray Eagle training takes place at Libby Army Airfield.

Expansion Potential

The IQT common core and the course for Shadow are currently running 24-hour operations. It would be difficult to expand the student capacity without committing more resources. The Gray Eagle course is just beginning, and the Hunter course is winding down. There is enough infrastructure for the planned capacity. Any expansion would require more resources.

Training and Training Facilities

Most UAS training operations take place at the Black Tower complex. Gray Eagle and Warrior Alpha vehicles are flown out of Libby Army Airfield.

Military and General Support Mission

Fort Huachuca is home to the military intelligence schoolhouse. This proximity could provide a unique opportunity for collective training between the two schoolhouses. Huachuca is also home to the Buffalo Soldier Electronic Testing Range.

Weather

Fort Huachuca has excellent weather for UAS operations. It has approximately 270 training days a year.

Location

See Figure C.3. Fort Huachuca is situated in southeastern Arizona, near the town of Sierra Vista. There are approximately 60,000 people at Fort Huachuca.

Figure C.2
Airfield and Maintenance Facilities at Fort Huachuca

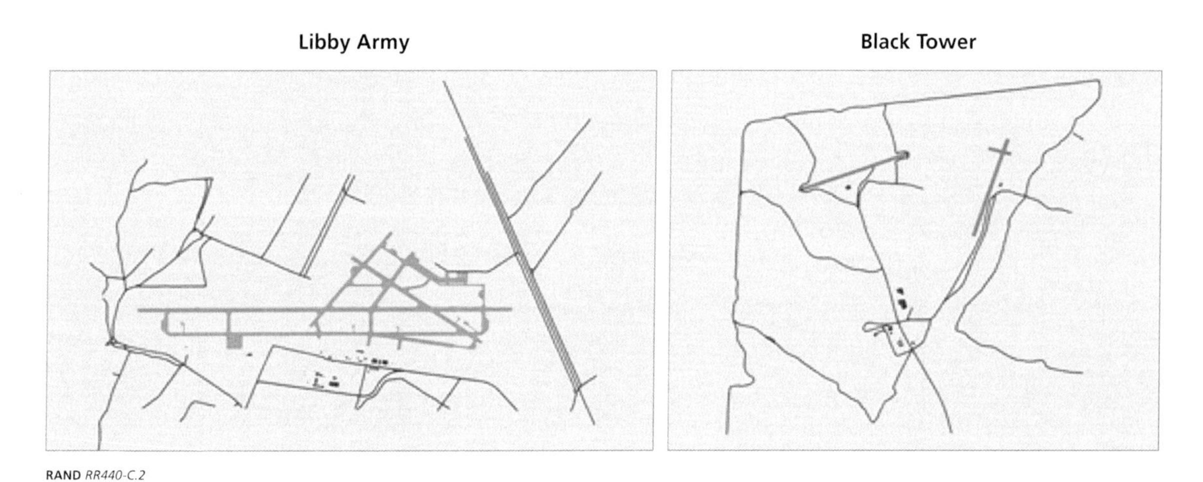

RAND *RR440-C.2*

Figure C.3
Location of Fort Huachuca

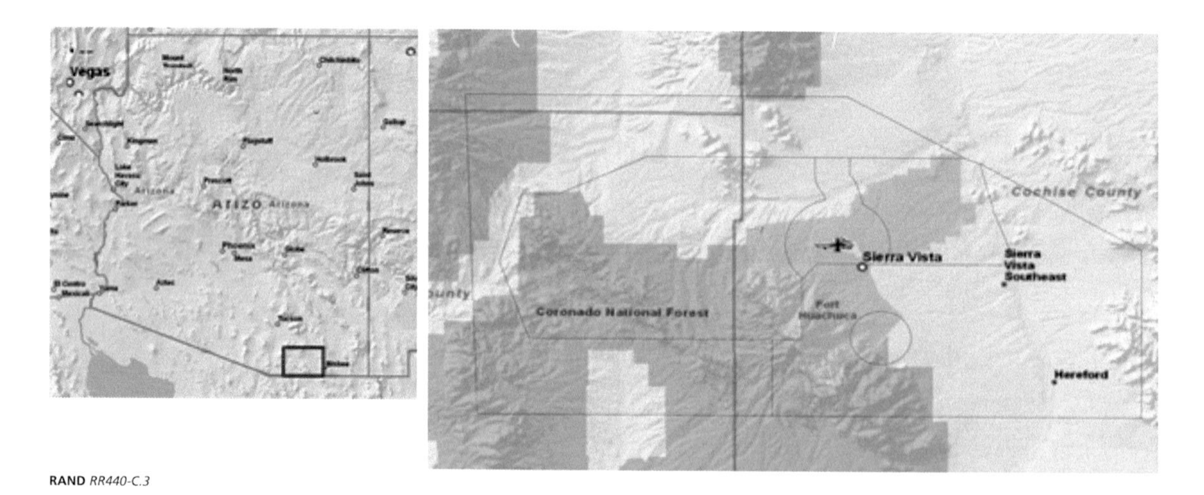

RAND *RR440-C.3*

Base Loading

Fort Huachuca is the hub for Army UAS training. This consolidation could be both good and bad. The lack of redundancy elsewhere limits the ability to expand operations when needed. For example, if one of the runways were damaged, the entire Army UAS training capability could be severely affected.

Air Force Active Duty MQ-1/9 Formal Training Unit—Holloman AFB, New Mexico

Flight Training Areas and Airspace

See Figure C.4. Holloman AFB has access to the restricted airspace areas within the White Sands Missile Range (WSMR) and McGregor airspace. It is possible to reach both ranges without leaving restricted airspace. While competition for airspace is fierce, WSMR supports flying RPA operations out of Holloman AFB. The current plan is for MQ1/MQ9 to depart Holloman AFB northbound for WSMR airspace and to remain totally within R-5107 B/C/D/H. An alternative is to depart Holloman AFB southbound and climb to FL-180 in R-5107D. There is a three-way memorandum of understanding in place, but the testing mission at WSMR occasionally conflicts with RPA training. A few other areas near Holloman AFB have potential for RPA operations, although they require COAs from the FAA.

The MQ-1/9 FTU completed the transition from Creech AFB to Holloman AFB. Plans call for an additional FTU, bringing a total of five MQ-1/9 training squadrons, including one maintenance squadron, to Holloman AFB. These squadrons will fall under the 49th Wing and bring in an additional 200 officers, 250 enlisted, and 150 contractors. Two hundred students are expected to cycle through in three-month training periods. Holloman AFB would have 28 MQ-1s and 10 MQ-9s. To operate the aircraft, they would have 12 common ground stations, four Predator Primary Satellite Links, two LREs, and five mission control elements. They expect to fly approximately 2,800 sorties a year, 540 at night.

Figure C.4
Restricted and Non–Joint Use Class D Airspace near Holloman AFB

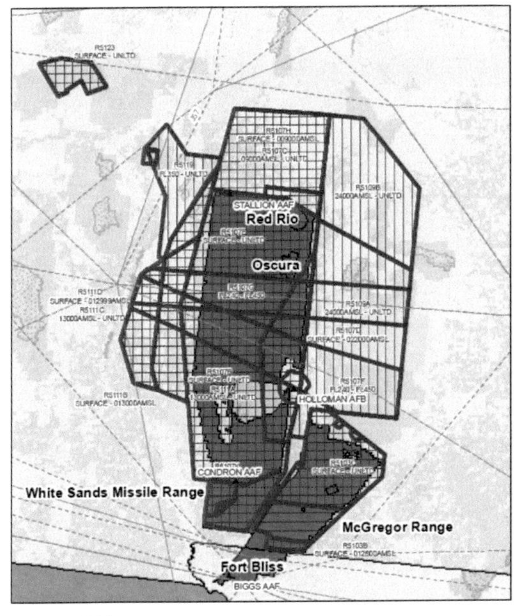

	Altitudes		Hours of Use	
	Minimum	Maximum	From	To
Airspace				
Beak A MOA	12,500 feet MSL	UTBNI FL180	0600	Sunset
Beak B MOA	12,500 feet MSL	UTBNI FL180	0600	Sunset
Beak C MOA	12,500 feet MSL	UTBNI FL180	0600	Sunset
Beak A/B/C ATCAAs	FL180	UTBNI FL230	As scheduled and coordinated	
Ancho A/B/C ATCAAs	FL180	UTBNI FL230	As scheduled and coordinated	
Cowboy A/B/C ATCAAs	FL230	FL600	As scheduled and coordinated	
Talon High East MOA	12,500 feet MSL	FL180	Sunrise	Sunset
Talon High West MOA	12,500 feet MSL	FL180	Sunrise	Sunset
Talow Low MOA	300 FEET AGL	UTBNI 12,500 feet MSL	Sunrise	Sunset
Talon High East and West ATCAAs	FL180	FL600	As scheduled and coordinated	
Valmont ATCAA	FL180	FL600	As scheduled and coordinated	
Restricted Area				
R-5103B (McGregor)	Surface	Unlimited	0700†	2000†
R-5103C (McGregor)	Surface	Unlimited	0700†	2000†
R-5107A (Ft Bliss)	Surface	Unlimited	Continuous	
R-5107B (WSMR)	Surface	Unlimited	Continuous	
R-5107C (WSMR)	9,000 feet MSL	Unlimited	Continuous Monday - Friday 12 hours in advance††	
R-5107D (WSMR)	Surface	FL220	Continuous	
R-5107E (WSMR)	Surface	Unlimited	By NoTAM 12 hours in advance	
R-5107F (WSMR)	FL240	FL450	Continuous M-F; 12 hours in advance pm weekends	
R-5107G (WSMR)	FL240	FL450	Continuous M-F; 12 hours in advance pm weekends	
R-5107H (WSMR)	Surface	UTBNI 9,000 feet MSL	By NoTAM 12 hours in advance	
R-5107J (WSMR)	Surface	UTBNI 9,000 feet MSL	Continuous Mon-Fri 12 hours in advance ††	
R-5109A	24,000 feet MSL	Unlimited	Intermittent by NoTAM	
R-5109B	24,000 feet MSL	Unlimited	Intermittent by NoTAM	
R-5111A (WSMR)	13,000 feel MSL	Unlimited	By NoTAM 12 hours in advance	
R-5111B (WSMR)	Surface	UTBNI 13,000 feet MSL	By NoTAM 12 hours in advance	
R-5111C (WSMR)	13,000 feel MSL	Unlimited	By NoTAM 12 hours in advance	
R-5111D (WSMR)	Surface	13,000 feet MSL	By NoTAM 12 hours in advance	

Notes: † – other times by NOTAM, †† – other times by NOTAM, 12 hours in advance
UTBNI = Up to, but not including; AGL = above ground level; FL = Flight level. FL 180 is approximately 18,000
Feet MSL; MSL= mean sea level; NOTAM = Notice to Airmen
Source: U.S. Air Force 2006

RAND *RR440-C.4*

Airfield and Maintenance Facilities

These are included with the training facilities.

Expansion Potential

The second MQ-9 FTU is scheduled to open in FY 2013, making Holloman home to three MQ-1/MQ-9 FTUs. There is currently a year lag in infrastructure to be able to support a third FTU. The new facilities were supposed to have been finished prior to the standup of the new FTU; however, construction is running approximately a year behind schedule. It is difficult to comment on the ability to meet the required capacity as the capacity requirement is in flux.

Training and Training Facilities

See Figure C.5. Holloman AFB has 200,000 ft^2 of existing and open facilities, including office space and hangars that could accommodate the FTUs with minor renovations. The map in Figure C.5 shows where new facilities would be located and identifies existing facilities.

Table C.1 shows the space required to start up the unit, as well as the required final bed-down space, as defined by military construction. Table C.2 shows the facility plan for each functional requirement.

Figure C.5
MQ-1/MQ-9 Beddown Plan

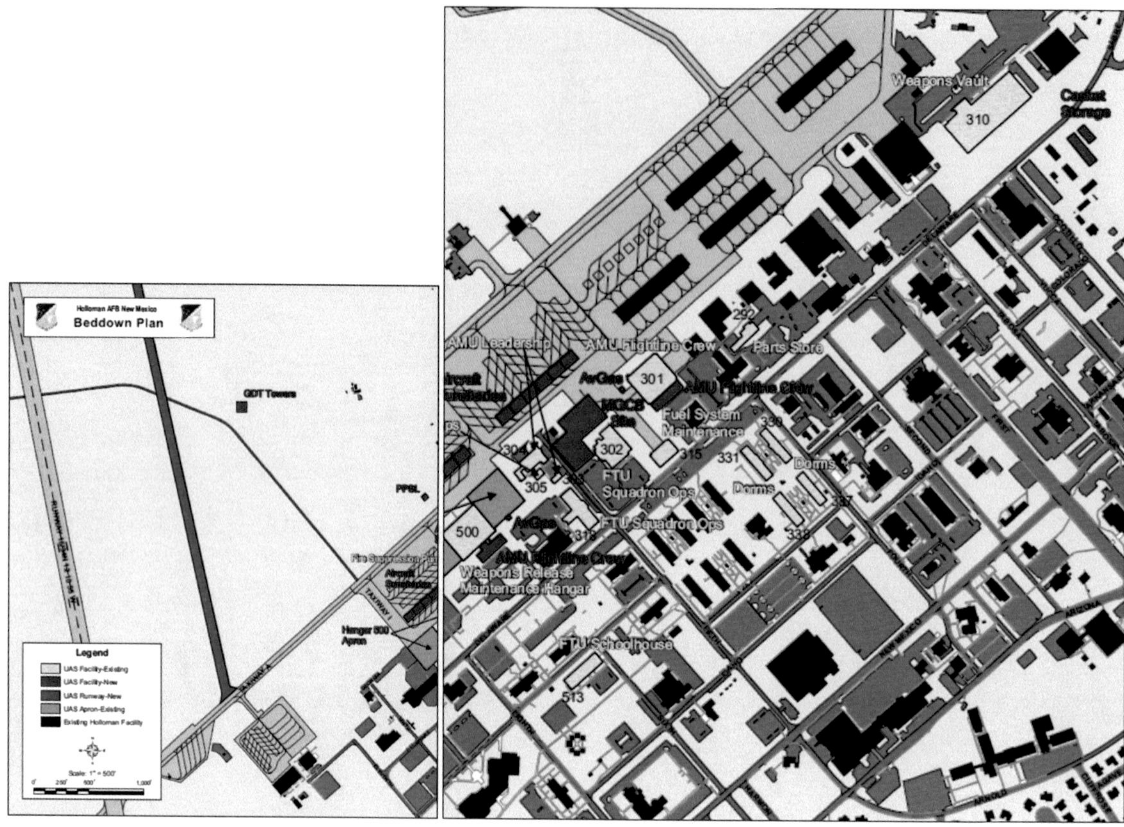

RAND *RR440-C.5*

Military and General Support Mission

Other aircraft flown at Holloman currently include the T-38, MQ-1, MQ-9, the QF-4, and the German Tornado.[3] There are plans to move at least a portion of the F-16 FTUs from Luke AFB, Arizona, to Holloman AFB. This could affect the availability of airspace access.

Location

Holloman AFB is located in southern New Mexico, near the town of Alamogordo. The base encompasses about 59,600 acres and has a population of about 21,000 military, civilian, and families.

Base Loading

Moving the FTUs from Creech AFB to Holloman AFB freed up some capacity at the former and allowed the latter to focus purely on training.[4] There is some redundancy with the Air National Guard FTU at March ARB; however, that FTU is much smaller and primarily trains

[3] A German training unit is stationed at Holloman AFB. The F-22s will be transitioning from the base. Air Combat Command and the Air Staff are deciding what to put in place of the F-22 squadron.

[4] The LRE training is a separate course from the FTU and still taught at Creech.

Table C.1
Beddown Plans

Function/Activity	Space Required for Initial Startup	Space Required for Final Beddown
Parking apron (000 ft^2)	30	60
Squadron operations facility (000 ft^2)	16	48
FTU schoolhouse	20	50
(classrooms and simulators) (000 ft^2)	11	11
MCE facility (000 ft^2)	12	24
Aircraft maintenance unit (000 ft^2)	30	70
Munitions PGM shop (000 ft^2)	2	Unknown
Munitions storage (000 ft^2)	3	Unknown
Aircraft parts store (000 ft^2)	10	10
Weapon load trainer (bays)	1 bay	1 bay[a]
Casket storage (000 ft^2)	8	16
Bulk fuel storage (gal)	32	32
Lodging (rooms)	60	200
All backshops (000 ft^2)	24	24

[a] Use maintenance bay.

Air National Guard operators. March ARB currently has adequate infrastructure and capacity for its training mission.

Air Force RQ-4 Formal Training Unit—Beale AFB, California

Flight Training Areas and Airspace

Because of the altitude at which RQ-4s conduct their primary mission, airspace access is manageable at Beale AFB. A cylinder is used to access Class A airspace. The FAA provides a standing window for flights into this cylinder of 10 hours a day, every weekday. Requests outside this window are handled individually and have not been a limitation.

Airfield and Maintenance Facilities

Beale AFB has dedicated airfield and maintenance facilities.

Expansion Potential

There is enough capacity for expanding the training mission of RQ-4s at Beale AFB. There are plans to include some portion of the operator or maintenance training for the Navy's MQ-4 BAMS at Beale, but this would be several years from now.

Table C.2
Facility Requirements

Function Description	Remarks
Flightline pavement	Use main ramp.
Live ordnance load area	Construct new LOLA on taxiway Echo.
Maintenance hangar (new FTU)	Use building 500.
Maintenance hangar (3 FTU squadrons)	Use building 301.
FTU squadron operations (new FTU)	For initial capability, use building 513 and a portion of building 302; when project is complete, transition from building 513 into building 318.
FTU squadron operations (Creech UAS)	If there are two FTU squadrons, use building 318; a third FTY squadron can occupy all of building 302 after F-22A transition is complete.
Aircraft maintenance unit (new FTU)	Initially locate leadership team in building 303 and operate flightline crews out of building 301 until completion of building 500 new construction (10,000 ft^2).
Aircraft maintenance unit (3 FTU squadrons)	If two units, locate both leadership teams in building 303 and flightline crews of the second unit in building 301. If three units, locate third leadership team in building 302 and locate flightline crew in building 302. New construction for building 301 can also be considered for the third crew.
Fuel system maintenance	Use building 315.
Precision guided munitions facility	Construct two maintenance bays and administration.
Munitions storage	Construct 26-by-120-ft Hayman igloo (possibly two 60-ft sections).
Aircraft parts store (new FTU)	Use existing contract support or shared building 292 (T-38 parts store).
Aircraft parts store (3 FTU squadrons)	If T-38 mission relocates, use building 292. If no relocation, add space to building 292.
Weapon release shop	Use each respective maintenance bay (building 500 for one and building 302 for two).
Casket storage	Construct 50-by-800-ft covered storage pad in logistics readiness squadron yard. Requirement may grow, depending on quantity of MQ-9 caskets on hand.
Bulk fuel storage	Construct two 8,000-gallon tanks adjacent to hangars 301 and 500 (for aviation gas) for MQ-1 and use existing JP-8 capacity for MQ-9.
Various backshops	Construct 5,000-ft^2 addition on building 500. Building 301 may require new additional space.

Training and Training Facilities

Beale AFB has adequate training space. There are several different aspects of training: classroom, simulation, and live flights. There is enough classroom space available for both the sensor operator training and the pilot training. The pilots use a pilot simulator to cover emergency procedures and aircraft operation. This simulator is very limited in its ability to mimic real-life contingencies. The sensor operators train on Data Analysis Workstations. These two simulators are not compatible, so there is no true crew simulator. Such a simulator has been

Figure C.6
Location of Holloman AFB

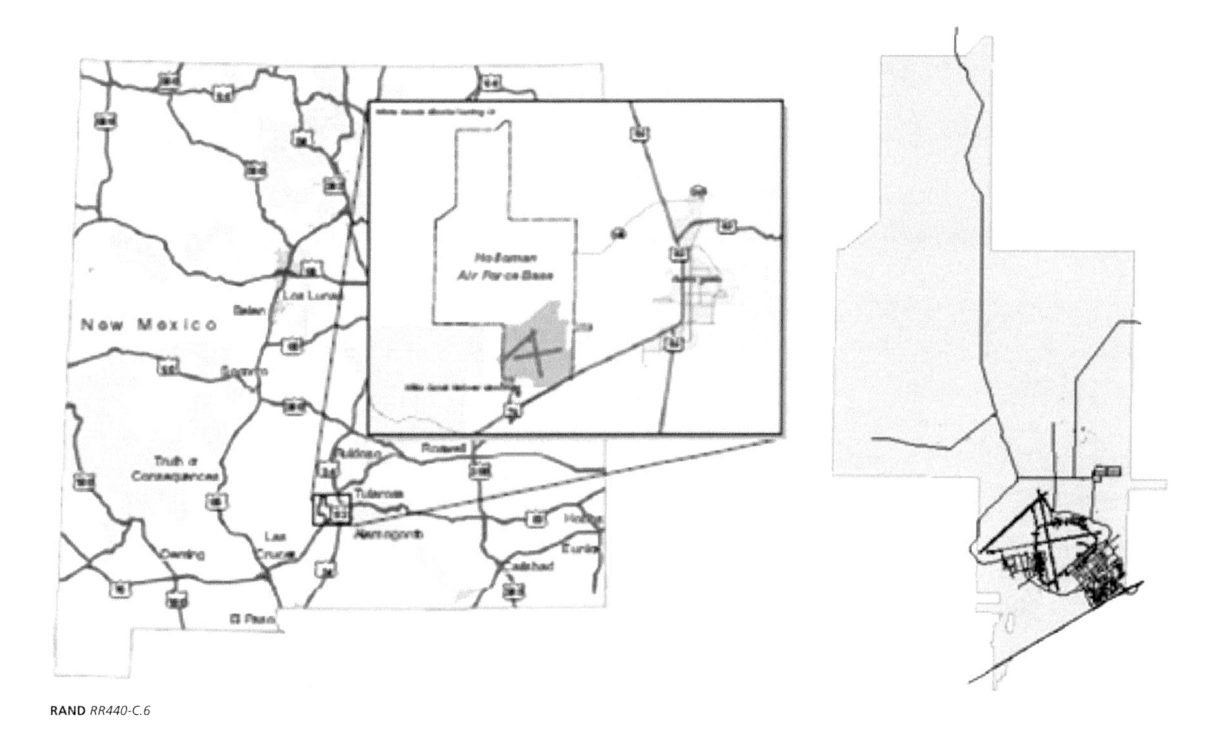

RAND *RR440-C.6*

programmed but is still eight to ten years away.[5] Crew training takes place on live missions, with instructors flying the mission with students.

Military and General Support Mission

Beale AFB is home to several other aircraft, including the MC-12, the U-2, and the T-38. The RQ-4 fits well with the mission of other aircraft at Beale since it was designed to eventually replace the U-2. RQ-4 pilots are also doing tours in the MC-12. There was some talk of maintaining dual currency in the two platforms, but the status of that plan is unknown.

Location

Beale AFB is located in northern California, approximately one hour outside Sacramento. The nearest town is Marysville, California.

Base Loading

Beale AFB is the hub for high-altitude surveillance. The installation is extremely large, with plenty of room for growth.

---

[5] We understand this decision to be based on resources since on-the-job training on live missions is available.

References

Ainsworth, Bernadette L., "Mini-Plane Newest Addition to Unmanned Family," Transformation website, U.S. Department of Defense, October 17, 2005.

Anderson, Michael G., "The Air Force Rapid Response Process: Streamlined Acquisition During Operations Desert Shield and Desert Storm," Santa Monica, Calif.: RAND Corporation, N-3610/3-AF, 1992. As of March 3, 2014:
http://www.rand.org/pubs/notes/N3610z3.html

Association of the United States Army, *Rapid Equipping Force: Innovative Materiel Solutions to Operational Requirements*, 2003.

Army Tactics, Techniques and Procedures (ATTP) 3-04.15 (Field Manual 3-04.15), Marine Corps Reference Publication (MCRP) 3-42.1A, Navy Tactics, Techniques and Procedures (NTTP) 3-55.14, and Air Force Tactics, Techniques and Procedures (AFTTP) 3-2.64, *Multi-Service Tactics, Techniques, and Procedures for Unmanned Aircraft Systems*, Joint Base Langley-Eustis, Va.: Air Land Seal Application Center, September 2011.

Baldor, Lolita C., "Military to Develop More Sophisticated Drones," *Fayetteville Observer*, November 5, 2010.

Bennett, John T., "Gates' ISR Task Force to Join Top DoD Intel Office," *Defense News*, October 7, 2010.

Best, Jr., Richard A., "Intelligence, Surveillance, and Reconnaissance (ISR) Acquisition: Issues for Congress," Washington, D.C.: Congressional Research Service, 2010.

Blickstein, Irv, and Charles Nemfakos, "Improving Acquisition Outcomes: Organizational and Management Issues," Santa Monica, Calif.: RAND Corporation, OP-262-OSD, 2009. As of March 3, 2014:
http://www.rand.org/pubs/occasional_papers/OP262.html

Buchanan, David R., *Joint Doctrine for Unmanned Aircraft Systems: The Air Force and the Army Hold the Key to Success*, Newport, R.I.: Naval War College, 2010.

Chairman, House Armed Services Committee, "H.R. 4310—FY13 National Defense Authorization Bill Chairman's Mark," undated. As of May 2, 2014:
http://armedservices.house.gov/index.cfm/files/serve?File_id=e7c34102-53e4-455a-b345-358f3e99e8cc

Chairman of the Joint Chiefs of Staff Guide 3501, "The Joint Training System: A Guide for Senior Leaders," June 8, 2012.

Chairman of the Joint Chiefs of Staff Instruction 3500.01G, "Joint Training Policy and Guidance for the Armed Forces of the United States," March 15, 2012.

Chairman of the Joint Chiefs Instruction 8501.01A, Chairman of the Joint Chiefs of Staff, Combatant Commanders, and Joint Staff Participation in the Planning, Programming, Budgeting, and Execution System, December 3, 2004, current as of February 12, 2008.

Christensen, Clayton M., and Michael Overdorf, "Meeting the Challenge of Disruptive Change," *Harvard Business Review*, March 2000.

Congressional Budget Office, *Policy Options for Unmanned Aircraft Systems*, Washington, D.C., Pub. No. 4083, June 9, 2011.

Defense Science Board, *Report of the Task Force on Training Superiority and Training Surprise*, Washington, D.C.: Office of the Under Secretary of Defense for Acquisition, Technology, and Logistics, January 2001.

———, *Report of the Task Force on Training for Future Conflicts: Final Report*, Washington, D.C.: Office of the Under Secretary of Defense for Acquisition, Technology, and Logistics, June 2003.

———, *Report of the Task Force on Fulfillment of Urgent Operational Needs*, Washington D.C.: Office of the Under Secretary of Defense for Acquisition, Technology, and Logistics, 2009.

Department of the Navy, *Navy Planning Guide 2010*, 2010. As of May 1, 2014:
http://www.navy.mil/navydata/policy/seapower/sne10/sne10-all.pdf

Dietrich, Shane, *Wartime Test and Evaluation: Initiatives Lead to Cultural Change*, Carlisle Barracks, Pa.: U.S. Army War College, 2007.

Director, Readiness and Training Policy and Programs, *Department of Defense Training Transformation Implementation Plan FY2006–FY2011*, Washington, D.C.: Office of the Under Secretary of Defense for Personnel and Readiness, February 23, 2006

Drezner, Jeffrey A., Geoffrey Sommer, and Robert S. Leonard, *Innovative Management in the DARPA High Altitude Endurance Unmanned Aerial Vehicle Program: Phase II Experience*, Santa Monica, Calif.: RAND Corporation, MR-1054-DARPA, 1999. As of March 3, 2014:
http://www.rand.org/pubs/monographs/MG350.html

DSB—*See* Defense Science Board.

Erwin, Sandra I., "To Meet Urgent Needs, Commanders Bypass Pentagon Acquisition System," *National Defense*, July 2010.

Francis, Paul, "Defense Acquisitions: Charting a Course for Lasting Reform," Testimony Before the Committee on Armed Services, U.S. House of Representatives, Washington, D.C.: Government Accountability Office, GAO-09-663T, April 30, 2009.

GAO—*See* Government Accountability Office.

Gates, Robert, Secretary Gates' Speech at National Defense University, September 29, 2008.

Gertler, Jeremiah, *U.S. Unmanned Aerial Systems*, Washington, D.C.: Congressional Research Service, R42136, January 3, 2012.

Government Accountability Office, Army Modernization: The Warfighting Rapid Acquisition Program Needs More Specific Guidance," Washington, D.C., GAO/NSIAD-99-11, November 1998.

———, "Military Training: Actions Needed to Enhance DOD's Program to Transform Joint Training," Washington, D.C., GAO-05-548, June 2005.

———, "Defense Acquisitions: Status and Challenges of Joint Forces Command's Limited Acquisition Authority," Washington, D.C., GAO-07-546, April 2007a.

———, "Defense Acquisitions: An Analysis of the Special Operations Command's Management of Weapon System Programs," Washington, D.C., GAO-07-620, June 2007b.

———, "Unmanned Aircraft Systems: Federal Actions Needed to Ensure Safety and Expand Their Potential Uses within the National Airspace System," Washington, D.C., GAO-08-511, May 2008a.

———, "Unmanned Aircraft Systems: Additional Actions Needed to Improve Management and Integration of DOD Efforts to Support Warfighter Needs," Washington, D.C., GAO-09-175, November 2008b.

———, "Unmanned Aircraft Systems: Comprehensive Planning and a Results-Oriented Training Strategy Are Needed to Support Growing Inventories," Washington, D.C., GAO-10-331, March 2010a.

———, "Warfighter Support: Improvements to DOD's Urgent Needs Processes Would Enhance Oversight and Expedite Efforts to Meet Critical Warfighter Needs," Washington, D.C., GAO-10-460, April 2010b.

Govindarajan, Vijay, Praveen K. Kapalle, and Erwin Daneels, "The Effects of Mainstream and Emerging Customer Orientations on Radical and Disruptive Innovations," *Journal of Product Innovation Management*, Vol. 28, No. 1, November 2011.

Henderson, Rebecca M., and Kim B. Clark, "Architectural Innovation: The Reconfiguration of Existing Product Technologies and the Failure of Existing Firms," *Administrative Science Quarterly*, Vol. 35, No. 1, March 1990.

Hicks, Kathleen H., *Invigorating Defense Governance: a Beyond Goldwater-Nichols Phase 4 Report*, Washington, D.C.: Center for Strategic and International Studies, 2008.

Hicks, Kathleen H., David Berteau, Samuel J. Brannen, Eleanore Douglas, Nathan Freier, Clark A. Murdock, and Christine E. Wormuth, *Transitioning Defense Organizational Initiatives: An Assessment of Key 2001–2008 Defense Reforms*, Washington, D.C.: Center for Strategic and International Studies, December 2008.

House Armed Services Committee, *Conference Report Accompanying the Fiscal Year 2013 National Defense Appropriations Act*, 112th Cong., 2nd Sess., December 17, 2012. As of March 28, 2014: http://docs.house.gov/billsthisweek/20121217/CRPT-112HRPT-705.pdf

Joint Publication 1-02, *Department of Defense Dictionary of Military and Associated Terms*, Washington, D.C.: Joint Chiefs of Staff, 2013.

——— 3-0, *Doctrine for Joint Operations*, Washington, D.C.: Joint Chiefs of Staff, 2011.

——— 3-55.1, *Joint Tactics, Techniques, and Procedures for Unmanned Aerial Vehicles*, Washington, D.C.: Joint Chiefs of Staff, 1993.

Kennedy, Harold, "Army 'Rapid Equipping Force' Taking Root," *National Defense,* Vol. XCI, No. 635, October 2006.

Kennedy, Tim, "TRADOC Seeks Wartime Solutions from Rapid Equipping Force," *Army*, Vol. 54, No. 8, August 2004.

te Kulve, Haico, and Wim A. Smit, "Novel Naval Technologies: Sustaining or Disrupting Naval Doctrine," *Technological Forecasting and Social Change,* Vol. 77, No. 7, September 2010.

Looy, Bart Van, Thierry Martens, and Koenraad Debackere, "Organizing for Continuous Innovation: On the Sustainability of Ambidexterous Organization," *Creativity and Innovation Management,* Vol. 14, No. 3, August 31, 2005.

Loxterkamp, Ed, "ISR Task Force Areas of Interest to EUCOM and AFRICOM: ISR Task Force Rapid Acquisition IPT," briefing, U.S. European Command and U.S. Africa Command Science Technology Conference, 14–18 June, 2010.

McLeary, Paul, "Industry, U.S. Army Pushing for More Helicopter Teaming," *Defense News*, December 15, 2012.

Middleton, Michael W., *Assessing the Value of the Joint Rapid Acquisition Cell,* master's thesis, Monterey, Calif.: Naval Postgraduate School, 2006.

Miles, Donna, "Rapid Equipping Force Speeds New Technology to Front Lines," American Forces Press Service, August 12, 2005.

MTTP—*See* Army Tactics, Techniques and Procedures . . . .

Murdock, Clark A., and Michele A. Flournoy, *Beyond Goldwater-Nichols: U.S. Government and Defense Reform for a New Strategic Era: Phase 2 Report*, Washington, D.C.: Center for Strategic and International Studies, 2005.

Pickup, Sharon, *Air Force Training: Actions Needed to Better Manage and Determine Costs of Virtual Training Efforts*, Washington, D.C.: Government Accountability Office, July 19, 2012.

Schank, John F., Harry J. Thie, Clifford M. Graf II, Joseph Beel, and Jerry M. Sollinger, *Finding the Right Balance: Simulator and Live Training for Navy Units*, Santa Monica, Calif.: RAND Corporation, MR-1441-NAVY, 2002. March 3, 2014: http://www.rand.org/pubs/monograph_reports/MR1441.html

Sherman, Jason, "DoD Approves 'Surge' of ISR Personnel, Seeks $1 Billion More in FY-09," *Inside the Air Force*, 2008.

Stavridis, James G., *Partnership for the Americas: Western Hemisphere Strategy and U.S. Southern Command*, Washington, D.C.: National Defense University Press, 2010.

Sullivan, Michael J., "Rapid Acquisition of Mine Resistant Ambush Protected Vehicles," letter, GAO-08-884R, Washington, D.C., July 2008.

———, "Defense Acquisitions: Perspectives on Potential Changes to Department of Defense Acquisition Management Framework," letter to congressional committees, Washington, D.C., GAO-09-295R, February 27, 2009a.

———, "Defense Acquisitions: Rapid Acquisition of MRAP Vehicles," Testimony Before the House Armed Services Committee, Defense Acquisition Reform Panel, Washington, D.C.: Government Accountability Office, GAO-10-155T, October 2009b.

Tellis, Gerard J., "Disruptive Technology or Visionary Leadership?" *Journal of Product Innovation Management,* Vol. 23, No. 1, January 2006.

Thompson, Mark, "Costly Flight Hours," *Time*, April 2, 2013. As of March 26, 2014:
http://nation.time.com/2013/04/02/costly-flight-hours/print/

UAS Task Force Airspace Integration Product Team, "Unmanned Air Craft Systems Airspace Integration Plan," Washington, D.C.: U.S. Department of Defense, March 2011.

U.S. Army Maneuver Center of Excellence, "Small UAS (Raven) Master Trainer /4D-F8/600-F20," course description, undated. As of May 6, 2013:
http://www.benning.army.mil/infantry/197th/229/SUASMT/

U.S. Air Force, "Report on Future Unmanned Aerial Systems Training, Operations, and Sustainability," 2011.

———, *RPA Vector: Vision and Enabling Concepts, 2013–2038*, Washington, D.C., Headquarters, United States Air Force, February 17, 2014.

Vergakis, Brock, "Navy Completes 1st Unmanned Carrier Landing," Associated Press, July 10, 2013. As of May 1, 2013:
http://bigstory.ap.org/article/navy-attempt-1st-unmanned-carrier-landing